Association for Women in Mathematics Series

Volume 36

Series Editor

Kristin Lauter, Facebook, Seattle, USA

Focusing on the groundbreaking work of women in mathematics past, present, and future, Springer's Association for Women in Mathematics Series presents the latest research and proceedings of conferences worldwide organized by the Association for Women in Mathematics (AWM). All works are peer-reviewed to meet the highest standards of scientific literature, while presenting topics at the cutting edge of pure and applied mathematics, as well as in the areas of mathematical education and history. Since its inception in 1971, The Association for Women in Mathematics has been a non-profit organization designed to help encourage women and girls to study and pursue active careers in mathematics and the mathematical sciences and to promote equal opportunity and equal treatment of women and girls in the mathematical sciences. Currently, the organization represents more than 3000 members and 200 institutions constituting a broad spectrum of the mathematical community in the United States and around the world.

Titles from this series are indexed by Scopus.

Abra Brisbin • Karen Lange • Erin McNicholas •
Emilie Purvine
Editors

Research Connections

Career and Research Journeys from the SMP Community

Springer

Editors
Abra Brisbin
Department of Mathematics
University of Wisconsin Eau Claire
Eau Claire, WI, USA

Karen Lange
Department of Mathematics
Wellesley College
Wellesley, MA, USA

Erin McNicholas
Mathematics
Willamette University
Salem, OR, USA

Emilie Purvine
Pacific Northwest National Laboratory
Seattle, WA, USA

ISSN 2364-5733 ISSN 2364-5741 (electronic)
Association for Women in Mathematics Series
ISBN 978-3-031-81219-4 ISBN 978-3-031-81217-0 (eBook)
https://doi.org/10.1007/978-3-031-81217-0

This Springer imprint is published by the registered company Springer Nature Switzerland AG
The registered company address is: Gewerbestrasse 11, 6330 Cham, Switzerland

We dedicate this volume to Deanna Haunsperger and Stephen Kennedy, and to the many instructors, teaching assistants, students, and visiting mathematicians who made the Carleton Summer Mathematics Program for Women (SMP) a transformational experience for us the editors, and for so many other participants. By organizing this book to mirror the structure of SMP, integrating personal connections with mathematical inquiry and discovery, we hope to honor the mathematical curiosity, mentorship, and community that Deanna and Steve's efforts have instilled.

Preface

Structure and Purpose of Research Connections

Is math research for me? Every undergraduate math major, and every first- and second-year math graduate student, has asked themselves this question at some point. Wrapped in this question are several more specific ones:

- What does math research really look like? And can *I* do it?
- I've always heard about subfield X, what's it all about?
- How do I get from here to there—what introductory material do I need to know in this subfield?
- Where do I go from there? What career paths have other mathematicians followed? What advice do they have for me?

And, especially for women and others from underrepresented groups in math,

- Do people like me go to graduate school, and succeed?

This book aims to provide perspectives on these important questions. Each chapter gives readers a taste of a different mathematician's work, presented with enough background material that an advanced undergraduate or early graduate student can understand the key ideas of the research. For example, Emily Olson introduces partially ordered sets and her work within a research community on a long-standing open conjecture one special case at a time. This belies the common impression of mathematics from textbooks and popular accounts, in which major theorems are typically presented fully-formed. Melinda Lanius provides a beautifully illustrated introduction to differential geometry and discretizations. And Jennifer Paulhus introduces both group actions and Riemann surfaces in a way that an advanced undergraduate can grasp. Rebecca Swanson's research falls in the category of the scholarship of teaching and learning. In her contribution, she provides an example of how academic mathematicians may interweave their research with teaching and mentoring the next generation of mathematicians. This book also showcases the non-academic career of Laurel Paget-Seekins, who worked in data science at the

Massachusetts Bay Transportation Authority. While these chapters are not intended to be a comprehensive examination of graduate-level mathematics—in either depth or breadth—they provide students with a "sneak preview" of what math research in a variety of subfields *could* look like, along with a taste of some of the fundamental knowledge that is needed to pursue research in that subfield. This format provides readers a sense of how to get "from here to there"—what topics they would need to learn and what styles of thinking they would need to employ, to specialize in a given area.

Each mathematical contribution is prefaced by a short biography of the mathematician who wrote the chapter, to give the reader a connection to the author and provide examples of paths from undergraduate education, through graduate school and beyond. Through these biographies, readers can see the person behind the mathematics and better imagine their own path. Unlike other valuable resources like *101 Careers in Mathematics* or WeUseMath.org, the goal of this book is not a comprehensive set of "final destinations" that are possible with a degree in math. We emphasize that these biographies are paths, discussing the opportunities, constraints, and factors that influenced each mathematician along the way to their current career. Reading these biographies may give current students ideas of opportunities (such as internships or reading courses) to pursue during their own educational journeys. Many of the writers have included advice for current students, based on their own experiences of how to succeed in graduate school.

We hope that reading these stories of how mathematicians decided to attend graduate school, choose their mathematical subfields, and find their current positions will help current students gain perspective on their own experiences. For example, reading Rebecca Swanson's uncertainty about whether she was "good enough to study math" may help encourage undergraduates who are wondering if they, too, have what it takes to succeed in graduate school. Emily Olson's story of changing graduate schools to be with her spouse may reassure graduate students wondering if it is valid to allow personal factors to influence their paths in graduate school.

Supporting Underrepresented Groups Within Mathematics

Women in mathematics often face special challenges. According to the National Science Foundation, while women earned between 42% and 45% of bachelor's degrees in mathematics and statistics each year from 2006 to 2016, they earned only 28–31% of doctoral degrees in math and statistics during the same time period. Part of the disparity may be attributed to a lack of role models: according to the Conference Board of Mathematical Sciences survey, from 2005 to 2015, women comprised between 11.4% and 16.0% of tenured and tenure-track faculty at PhD-granting universities.

A lack of role models or mentors like them in mathematics may lead female students and those from other underrepresented groups to question whether they really belong in graduate school. This can exacerbate the impostor phenomenon that nearly every graduate student experiences at one point or another. A lack of mentors, or a fear of seeming too focused on personal or family issues, may inhibit students from seeking practical advice, such as how to apply to graduate school or how to resolve a two-body problem. While this book focuses on the career paths and accomplishments of mathematicians from one program for women, the Carleton Summer Math Program, we believe that the lessons herein are valuable for all math students.

The Carleton College Summer Math Program for Women (SMP) was one of several programs in the late 1990s and early 2000s to support women in mathematics, including the Nebraska Conference for Undergraduate Women in Mathematics, EDGE (Enhancing Diversity in Graduate Education), and the George Washington University Summer Program for Women in Math. All of these programs have proven transformational for many of the students who participated in them. SMP was a four-week summer program for early-undergraduate math students (having finished their first or second year of college). In the 18 years it ran, between 1995 and 2014, 329 participants from across the United States took advanced math courses, attended colloquia by guest researchers, and forged social and mathematical connections with their peers from other colleges and universities. Starting in 2005, the program featured a conference, called the "SMPosium," which allowed program alumnae with PhDs to share their research with each other and to provide mentorship to the undergraduate participants. From 2010 to 2017, alumnae with support of the program directors also organized a mentoring workshop (Graduate Education Mentoring or GEM) the day before the Joint Mathematics Meetings, in which PhD alumnae mentored former participants currently in graduate school. These events fostered connections between "generations" of SMP participants and reinforced the sense of SMP as an ongoing community. SMP owes its effectiveness to the relentless efforts and enthusiasm of its directors, Deanna Haunsperger and Stephen Kennedy, as well as to the engaging teaching and mentorship of the program's instructors. Erica Flapan, Karen Brucks, Margaret Robinson, and Pam Richardson (who was also an SMP participant in 1999) all contributed to SMP's community by teaching courses in multiple summers.

The chapters in this book highlight the importance of building a support structure; of breaking a tough research problem into achievable sub-problems; of believing in yourself; of overcoming obstacles, both mathematical and personal; of illustrating mathematics in order to build your understanding of a concept; of having fun. We hope that every student contemplating graduate studies in mathematics can see something of themselves in the biographies and career paths described in this book, and use the research contributions to gain a greater clarity about potential mathematical subfields. Overall, this book aims to show the reader examples of building a greater sense of belonging, or a greater connection to the broader mathematical community. As attested by the biographies in this book, programs such as SMP have provided a sense of community and belonging for many mathematicians, and mathematical content that prepared them for research. Therefore, we also hope that this volume serves to encourage the continuation and development (and funding!) of programs that further develop these types of communities for women and other members of underrepresented groups in mathematics.

How to Use This Book

To Faculty

You can use this book as a classroom resource to introduce students to research in specific subfields of mathematics. For example, this book can be a source of ideas and starting points for in-class projects, independent studies, and student talks as well as supplementary reading in courses. The profiles of early career mathematicians and statisticians at the beginning of each chapter could be used as an advising resource for students considering graduate school, or even to show students a diverse view of modern mathematicians in a "Math for Liberal Arts"-style course. The tales of supportive professors and advisors contained in these profiles could also serve as personal inspiration of ways to reach out to promising students—especially those from traditionally underrepresented groups—and encourage their progress in mathematics.

To Students

You can peruse the articles in this book to get a sense of what subfields of mathematics interest you. Don't feel you need to read the chapters in order or read all of any given article. It's also fine to pause your reading of a given chapter, and pick it up in a future semester when you have more background. Even if an author's specific contributions to a field are currently beyond your understanding, you may still benefit from reading the introductory material in a chapter and the author's biography. An article that interests you could serve as a starting point for an independent study, a presentation at a Pi Mu Epsilon or other conference, or a research project.

If you are considering, or currently in, graduate school, we encourage you to read the biographies of any or all the mathematicians featured here—not just those in your subfield. You may discover someone in a very different area of mathematics, but a similar personal situation. Reading their stories may give useful food for thought you can apply to your own career path, either now or in the future.

You Belong Here

One of our favorite of the AWM EvenQuads playing cards deck highlighting the contributions of women in mathematics is a card simply entitled YOU Belong Here! As that card says, "Whether you are just beginning to explore your interests in mathematics or are a long-time devotee of the discipline, you are a unique contributor to the mathematics community."

Our discipline can seem so abstract, so separate from human experience. The underlying logical patterns feel more discovered than created. And while this may be true, these patterns are translated into the language of mathematics by people with unique tastes and perspectives. Deductive logic may be at the heart of everything we do, but there is also individual creativity, insight, and aesthetic value. Our hope for this project is that by putting the mathematics next to the biographies of the people who discovered it, you, the reader, will see the field for the human endeavor that it is, and see yourself as part of that endeavor. But if you don't see yourself in any of the stories in these pages, know that you still belong here! There is a place for you, and everyone who wants to be part of this amazing worldwide collaboration that has been ongoing since humans first started to make sense of the world around them. If you choose to invest your time, creativity, and efforts, the field of mathematics will be better for it. Your contributions will become intertwined with those of the millions of others who have found solace, beauty, diversion, creativity, and self-expression in mathematics—including the people highlighted in this book.

Additional Reading

We encourage interested readers to check out these other resources for more on career paths, personal stories, data, and mathematical content.

- For more about career paths in mathematics, readers may wish to peruse *101 Careers in Mathematics* and WeUseMath.org.
- For more personal stories of successes and struggles in math by a diverse set of mathematicians, we recommend *Living Proof: Stories of Resilience Along the Mathematical Journey*, which is available for free as an ebook from the Mathematical Association of America.
- For more holistic data about the representation of diverse groups in mathematics, readers can find the full National Science Foundation report *Women, Minorities, and Persons with Disabilities in Science and Engineering: 2019* at https://www.nsf.gov/statistics/wmpd. The Conference Board of Mathematical Sciences 2015 survey report is available at http://www.ams.org/profession/data/cbms-survey/cbms2015.
- For an account of the impact of the EDGE program, a program for women about to enter graduate school, readers are encouraged to read *A Celebration of the EDGE Program's Impact on the Mathematics Community and Beyond*, published by the Association for Women in Mathematics and Springer.

- Finally, many authors in this book have also provided references to excellent introductions to their fields. If a particular chapter piques your interest, please check out that chapter's bibliography for more reading on that subfield of math.

Eau Claire, WI, USA — Abra Brisbin
Wellesley, MA, USA — Karen Lange
Salem, OR, USA — Erin McNicholas
Seattle, WA, USA — Emilie Purvine
January 2024

Acknowledgments

This is the first time any of the four of us has served as an editor for a book, and we have learned a great deal along the way. We are grateful to the many people who have made this volume possible.

First and foremost, we would like to thank the authors of the chapters in this volume. Your elucidating writing about a wide variety of mathematical topics made this book a joy to edit. We are especially grateful for your patience as we navigated the world of academic book publishing, a process which was prolonged by the disruption of the COVID pandemic.

We would like to thank the anonymous reviewers who provided valuable feedback on the chapters in this volume, and helped us reach toward our goal of a book that would be accessible to upper-level undergraduates and first- and second-year math graduate students.

Thanks to the many wonderful people who helped this book come to fruition as part of Springer's Association for Women in Mathematics Series: Kristin Lauter, Dahlia Fisch, Nicholas DiBenedetto, Saveetha Balasundaram, Clement Wilson, Jeffrey Taub, and Bhuvana Ramakrishnan. Thank you for believing in our vision and for serving as guides along the way.

We thank Steve Kennedy, whose conversation with Karen and Erin in 2017 sparked the original idea for this book. Additional thanks to Cynthia Huffman for her encouragement and valuable suggestions that have shaped the final result.

Big thanks go out to all the members of the Carleton Summer Math Program (SMP) community: All the instructors, teaching assistants, and participants over the years have made SMP into more than a summer-long experience. SMP is a community that has shaped the careers and mathematical identities of many people, including us as editors, and the authors in this book. Particular thanks go to Deanna Haunsperger and Steve Kennedy, SMP's directors and visionary leaders, and to the instructors who helped shape SMP's welcoming and intellectually engaging culture throughout multiple summers: Erica Flapan, Karen Brucks, Margaret Robinson, and Pam Richardson.

Finally, thank you to our loved ones. Your ongoing support has made this book possible (even if some of you didn't really understand what we were working on).

With gratitude,

Abra Brisbin, Karen Lange, Erin McNicholas, and Emilie Purvine

Contents

Contributors

Abra Brisbin University of Wisconsin, Eau Claire, WI, USA

Karen Lange Wellesley College, Wellesley, MA, USA

Melinda Lanius Auburn University, Auburn, AL, USA

Erin McNicholas Willamette University, Salem, OR, USA

Claire Merriman George Mason University, Fairfax, VA, USA

Emily J. Olson Millikin University, Decatur, IL, USA

Laurel Paget-Seekins Massachusetts Bay Transportation Authority, Oakland, CA, USA

Jennifer Paulhus Grinnell College, Grinnell, IA, USA

Emilie Purvine Pacific Northwest National Laboratory, Seattle, WA, USA

Deena R. Schmidt University of Nevada, Reno, NV, USA

Elizabeth Stanhope Lewis & Clark College, Portland, OR, USA

Rebecca Swanson Colorado School of Mines, Golden, CO, USA

Chapter 1
On Partially Ordered Sets and the 1/3–2/3 Conjecture

Emily J. Olson

Abstract For this article, I assume that the reader has experience with proof writing but no other particular background. Since discrete mathematics courses vary drastically in terms of content, I include definitions that one would see in a full semester course on combinatorics or graph theory. We begin with background information on partially ordered sets. Let $(P, \leq)$ be a finite partially ordered set, where P has cardinality n. Consider linear extensions of P as permutations $x_1x_2 \cdots x_n$ in one-line notation. For distinct elements $x, y \in P$, we define $\mathbb{P}(x \prec y)$ to be the proportion of linear extensions of P in which x comes before y. For $0 \leq \alpha \leq \frac{1}{2}$, we say (x, y) is an α-balanced pair if $\alpha \leq \mathbb{P}(x \prec y) \leq 1 - \alpha$. The $1/3$–$2/3$ Conjecture states that every finite partially ordered set which is not a chain has a $1/3$-balanced pair. We include some history of the conjecture, including results commonly found in the literature. We make progress on this conjecture by showing that it holds for certain families of posets. These include lattices such as the Boolean, set partition, and subspace lattices, and partial orders that arise from Young diagrams. We also consider posets that satisfy the stronger condition of having a $1/2$-balanced pair. We pose various questions for future research throughout the article.

Keywords 1/3–2/3 Conjecture · α-balanced · Automorphism · Dimension · Lattice · Linear extension · Partially ordered set · Pattern avoidance · Poset · Width · Young diagram

1.1 Biography

When I graduated with my PhD, my biggest supporter gave me a binder he called *Emily's Book of Inspirational Women.* In it, he placed pictures of women that I admire along with a quote by each them. Some are scientists, like Katherine Johnson

E. J. Olson (✉)
Millikin University, Decatur, IL, USA
e-mail: ejolson@millikin.edu

A. Brisbin et al. (eds.), *Research Connections*, Association for Women in Mathematics Series 36, https://doi.org/10.1007/978-3-031-81217-0_1

and Sally Ride; others are modern, like Michelle Obama and Leslie Knope. As I paged through, blurry-eyed, I saw he choose Deanna Haunsperger as one of my inspirational women. There is no doubt that she belongs in my binder, and I am stunned that I have a relationship with someone who not only inspires me but also motivates me. I have turned the binder into a wall of inspirational women for my office, and Deanna will be there alongside the others. She and Steve Kennedy created a legacy that I am not only proud and honored to be a part of but also take responsibility to continue.

My journey into academia started long before my time in the Summer Mathematics Program (SMP) at Carleton College. Many of my friends in high school who liked STEM were going to major in premed in college. I was in good company when I applied to six colleges, and I briefly considered pursuing a premed major. I recall pondering over mathematics as a potential major and why it was not a more popular choice. I asked my AP Calculus teacher where would be a good school to study mathematics, and she told me the University of Wisconsin-Madison. Since I went to high school in Wisconsin, this school was an obvious choice for many in-state college-bound students. I did not want to do just what everyone else was doing, so I applied elsewhere.

After receiving a scholarship from Saint Louis University (SLU) in Saint Louis, Missouri, I decided to attend SLU for my undergraduate education. I had been denied admission to my top choices for college, so the scholarship was bittersweet. Making the most of the situation, I started college by taking Calculus II in the fall. Once I declared my major, I was assigned as an advisee to Dr. Christine Stevens, who turned into one of my influential mentors. Dr. Stevens saw my naive enthusiasm and guided me toward mathematically enriching summer activities. She advised that I apply for SMP in 2008, after my sophomore year. This program single-handedly changed my approach to the mathematics community. I found a group of people like me: curious students wanting to be convinced we should pursue mathematics. I recall working hard during my four weeks at SMP and also making friends that endure to this day. Deanna and Steve's program truly was my gateway into the community of mathematicians, a community of which I am proud to be a member.

The following summer I applied to, and was accepted into, a Research Experience for Undergraduates (REU). I chose to attend the REU at the Claremont Colleges. My group learned a substantial amount about p-adic numbers, but our project was fairly abstract, and I was frustrated with the lack of progress we made. We had one student in our group who already excelled at abstract thinking, so we relied on him for many of our small results. I know now that this is how math research goes sometimes, but I think it is unfortunate that we were not directed toward a more concrete project. I'll admit I was jealous that other groups at the REU made more progress than us. We presented a poster at the Joint Mathematics Meetings (JMM) in San Francisco in 2010. This was my first JMM, and so if nothing else, I participated in one of the most fundamental experiences in the life of a mathematician.

Two positive outcomes of my REU experience are the friends I made and the knowledge I gained about mathematical research. I decided to apply to graduate schools the following year. While I did not know what I would do with my PhD in Mathematics, SMP had exposed me to a myriad of options. I began graduate school at University of Nebraska-Lincoln (UNL), hoping to study group theory. I gave talks at seminars, took qualifying exams, and taught calculus recitations. However, my time at UNL was too short, as I decided to transfer to another program in order to be in the same state as my spouse, who was attending graduate school in physics. Before I changed schools, I sought out the advice of some of my SMP mentors, and one of them reminded me that I should do what is right for me. I felt very supported in making a choice.

Leaving UNL with a Master's degree, I attended Michigan State University for the remainder of my graduate studies. What made this transition difficult was that I had to start the PhD program from scratch, even though I had passed classes and qualifying exams at UNL. So, I took more classes and more rounds of exams. It was difficult to stay motivated during those extra years. However, I am not sure I would have survived graduate school without the support of my spouse, so I stand by my decision to move.

I was fortunate to be a teaching assistant for SMP during the summers of 2010, 2011, and 2013. These formative experiences deepened my connections with the SMP network and gave me a sense of what academia and teaching really entail. I found a passion for teaching and desired to pursue academia. I also gained insight into many different teaching styles. I am inspired by the awesome teaching skills of the SMP instructors. There is not one correct way to be a teacher, and I had to find what worked for me.

Extra motivation to finish graduate school came when I attended the SMP Graduate Education Mentoring (GEM) workshops run by SMP alumnae at JMM. These workshops focused on issues that occur in graduate school, allowed us to network with other SMPers, and gave us feedback on talks. I always left GEM a little more invigorated than when I came in. The extra years of graduate school wore me down, and I am incredibly grateful to all the SMPers who helped put my effort into perspective by joining me for coffee during a break or conversation at dinner.

I finally graduated in 2017. As an early career mathematician, I have looked ahead to see what I want to achieve. At a workshop recently, I was asked to create a 20-year plan and then work backward to see what I need to accomplish in 10 years, 5 years, and 3 years to reach these goals. When planning for the future, I think of my inspirational women. Truthfully, I am easily overwhelmed by what they have accomplished. I usually end up thinking that there is no way I can achieve those things. I probably thought the same thing when I started graduate school, and when I participated in the REU, and when I was in SMP. After all these years and through dedication, hard work, and lots of learning, I am now a faculty member who gets to research and teach mathematics. I have already achieved so much on my mathematical journey, and I hope to achieve so much more!

I would be completely justified putting every SMPer on my wall of inspirational women. We each bring our own important story, and I have often relied on the experiences of others to help me form my path forward. I am so grateful to everyone who has advised without criticism, listened without objection, and supported without judgment. Thank you all.

1.2 Introduction

1.2.1 Posets

We begin by defining the main combinatorial object of this paper: the partially ordered set. This cumbersome name is often shortened to "poset," which is the name most of the literature commonly uses. It is not surprising that while posets have a precise mathematical definition, most intuition for these objects comes from their pictorial representation. Both concepts are presented here.

Definition 1.1 A *partially ordered set*, or *poset*, is a set of elements P and binary operation $\leq$ that obey the following properties for all $x, y, z \in P$:

- Reflexive property: $x \leq x$,
- Antisymmetric property: if $x \leq y$ and $y \leq x$, then $x = y$,
- Transitive property: if $x \leq y$ and $y \leq z$, then $x \leq z$.

While this definition allows P to be an infinite set, we will assume here that all posets are finite.

Example 1.1 Consider a set with six elements, $\{a, b, c, d, e, f\}$. This set becomes a poset with the following relations:

$$\begin{array}{llllll} a \leq a & b \leq b & c \leq c & d \leq d & e \leq e & f \leq f \\ a \leq b & b \leq c & c \leq e & d \leq e & & \\ a \leq c & b \leq e & & d \leq f & & \\ a \leq d & & & & & \\ a \leq e & & & & & \\ a \leq f & & & & & \end{array}$$

We can check these relations satisfy the reflexive, antisymmetric, and transitive properties from Definition 1.1. For instance, we can see that $a \leq b$ and $b \leq e$. Therefore, it should be the case that $a \leq e$, and after checking the list, we see this is indeed true. This is an example of the transitive property. □

Next, let's draw this poset. We will draw one point for each element in the poset; in this case, we have six points. Now, for $x, y \in P$, we say x *covers* y, denoted $y \lessdot x$, if $y < x$ and no element $z \in P$ satisfies $y < z < x$. The elements x and y are *comparable* if either $x \leq y$ or $y \leq x$; otherwise, the pair is *incomparable*. We can create a graphical representation of a poset by drawing its cover relations. If we consider this picture as a graph, it is called the *Hasse diagram* of a finite poset P, whose vertices are elements of P and edges are cover relations, such that whenever

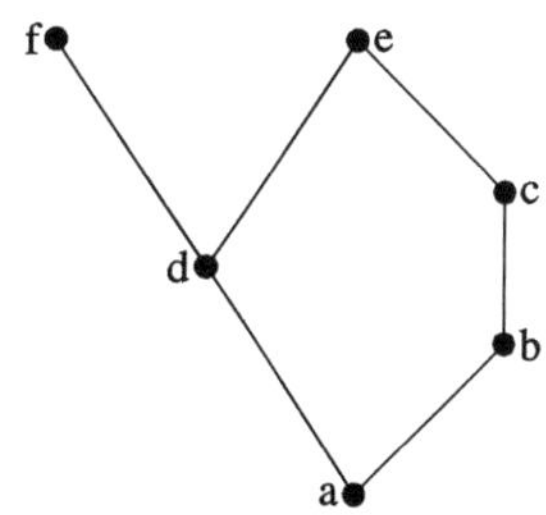

Fig. 1.1 A poset P with 6 elements

$x \lessdot y$, x is drawn below y. The Hasse diagram for Example 1.1 is drawn in Fig. 1.1. In practice, the Hasse diagram is all we need to define a poset; we almost never list out all the relations. As many mathematicians would say, "A picture is worth 1000 technical definitions."

Other terms that are used to describe parts of posets often make sense in context. For instance, an element $x \in P$ is *minimal* if for all y comparable to x, $y \geq x$. Similarly, $z \in P$ is *maximal* if for all y comparable to z, $y \leq z$. If a poset has only one minimal or maximal element, we call the element $\hat{0}$ or $\hat{1}$, respectively.

A subset of elements $S \subseteq P$ is called a *chain* if for all $u, v \in S$, either $u \leq v$ or $v \leq u$. The size of the largest chain in P is the *height* of P. Conversely, a subset of elements such that any two elements are incomparable in P is called an *antichain* and the size of the largest antichain of P is the *width* of P.

Example 1.2 Consider the poset given in Fig. 1.1. The subset $\{a, b, c, e\}$ is a chain in P, and $\{d, f\}$ is as well. Since $\{a, b, c, e\}$ is the largest chain in P, the poset P has height 4. An example of an antichain in P is $\{c, d\}$. Another antichain is $\{b, f\}$. In fact, there are no antichains with more than two elements, and so the width of P is 2. □

Throughout the text, we will denote a poset by $(P, \leq)$ or P if the binary relation is clear from context. It may occasionally be useful to distinguish the binary operation between posets, in which case we specify $\leq$ as $\leq_P$. By $x < y$, we mean that $x \leq y$ and $x \neq y$. The *dual* of P, denoted P^*, has the same elements as P and the partial order $x \leq_{P^*} y$ if $y \leq_P x$.

Our final definition in this section is order ideals. Some authors refer to these objects as "down sets" and "up sets," but we will call them lower and upper order ideals. These subsets of posets have many useful properties.

Definition 1.2 In a poset P, a *lower order ideal* is a subset $I \subset P$ such that if $x \in I$ and $y \leq x$, then $y \in I$. If there is some $x \in P$ which is the unique maximal element of the lower order ideal, we say the ideal is *principal*. We similarly define an *upper order ideal* and *principal upper order ideal generated by* x.

Example 1.3 Consider the poset given in Fig. 1.1. The subset $I = \{a, b, c, d\}$ is a lower order ideal of P. The subset $J = \{c, e, f\}$ is an upper order ideal. The principal lower order ideal generated by e is $\{a, b, c, d, e\}$. □

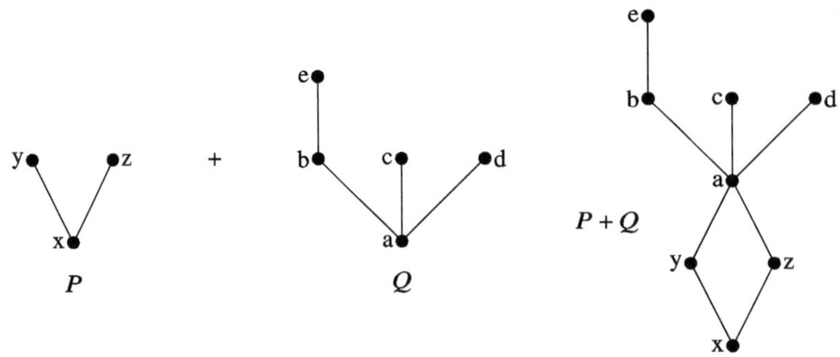

Fig. 1.2 Posets P and Q and their linear sum $P + Q$

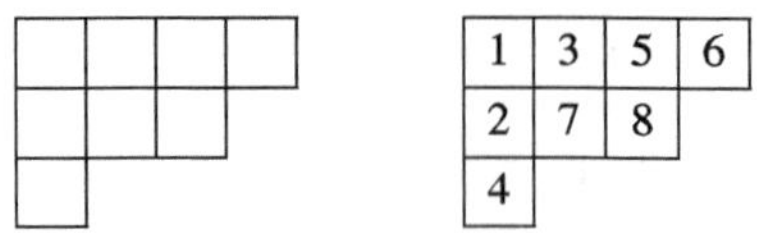

Fig. 1.3 A diagram of shape (4, 3, 1) and a standard Young tableau of the same shape

We can build new posets by combining two or more posets. One way to do this is through the linear sum. The *linear sum* of posets P and Q, denoted $P + Q$, is obtained by adding the relations $a < b$ for every $a \in P$ and $b \in Q$. In terms of the Hasse diagrams of P and Q, taking the linear sum $P + Q$ is the same as putting the Hasse diagram of Q above the Hasse diagram of P and drawing all covering relations between the maximal elements of P and the minimal elements of Q. See Fig. 1.2 for an example.

1.2.2 *Integer Partitions*

Another common and interesting combinatorial object is the integer partition. An *integer partition of n*, or *partition*, is a weakly decreasing sequence of positive integers $\lambda = (\lambda_1, \ldots, \lambda_k)$ that sums to n. To denote this, we use the notation $\lambda \vdash n$. For instance, if we let $n = 17$, then $\lambda = (9, 3, 3, 2)$ is an integer partition of n.

Definition 1.3 Let $\lambda = (\lambda_1, \ldots, \lambda_k)$ be a partition of n. The *Young diagram* corresponding to λ consists of k left-justified rows of cells where the ith row from the top has λ_i cells.

For an example of a Young diagram, see the left object in Fig. 1.3. We will always let n be the number of cells in a diagram, and since we often make no distinction between a partition and its Young diagram, the same notation for both suffices. We can also abbreviate by using exponential notation; we write $\lambda = (\lambda_1^{m_1}, \ldots, \lambda_k^{m_k})$ when the first m_1 rows of the diagram have length λ_1, etc. The cells of a Young diagram can be filled with integers from 1 to n using each number exactly once and increasing from left to right in the rows and top to bottom in the columns; this forms a *standard Young tableau*. For each Young diagram, there are many standard Young tableaux. We will refer to standard Young tableaux again in Sect. 1.5.4, so for now, it is just an interesting combinatorial object. An example of this can be seen on the right in Fig. 1.3.

As mathematicians like to do, we can generalize the notion of an integer partition and Young diagram. Perhaps the cells do not need to be left-aligned or perhaps we can remove the cells of a "smaller" Young diagram from a "bigger" Young diagram. These will be shifted diagrams and skew diagrams, respectively. What follows is a full definition.

Definition 1.4 If $\lambda \vdash n$ is a partition and if $\lambda_1 > \lambda_2 > \cdots > \lambda_k$, then λ is called a *strict partition*. The *shifted diagram* corresponding to a strict partition λ indents row i so that it begins on the diagonal cell (i, i). An example is given in Fig. 1.4a. A second type of diagram is a *skew diagram*, λ/μ, which is the set-theoretic difference between diagrams $\lambda = (\lambda_1, \ldots, \lambda_k)$ and $\mu = (\mu_1, \ldots, \mu_l)$ such that $\mu \subseteq \lambda$, that is, $l \leq k$ and $\mu_i \leq \lambda_i$ for each $1 \leq i \leq l$. A skew diagram can be either *left-justified*, as seen in Fig. 1.4b, or *shifted*, as seen in Fig. 1.4c.

We would like to point out that if we allow μ to be an empty partition, then λ/μ could refer to a Young, shifted, or skew diagram. To avoid ambiguity, we will always specify the type of diagram we intend, and if we refer to Young diagrams, we exclusively mean left-justified and non-skew diagrams.

For each diagram, there is a corresponding poset formed by letting each cell be an element of P. For $x, y \in P$, $x \leq y$ exactly when the cell for y is weakly to the right and/or weakly below the cell for x. See Fig. 1.5 for an example of the poset corresponding to the shifted diagram $(4, 2, 1)$. We will use the notation λ/μ to refer to a diagram and $P_{\lambda/\mu}$ for the corresponding poset.

1.2.3 Posets in the Wild: Directed Acyclic Graphs

If you have studied some graph theory, you will know that the placement of vertices of a graph on a page does not really matter. That is, there is no "up" and "down"

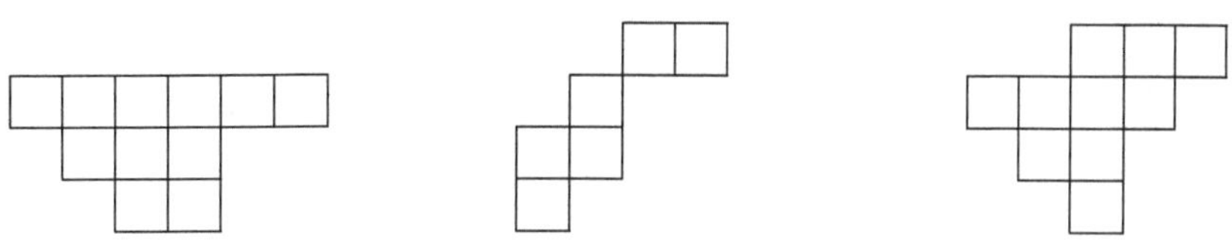

(a) (b) (c)

Fig. 1.4 (**a**) A shifted diagram of shape $(6, 3, 2)$, (**b**) a skew left-justified diagram of shape $(4, 2^2, 1)/(2, 1)$, and (**c**) a shifted skew diagram of shape $(6, 4, 2, 1)/(3)$

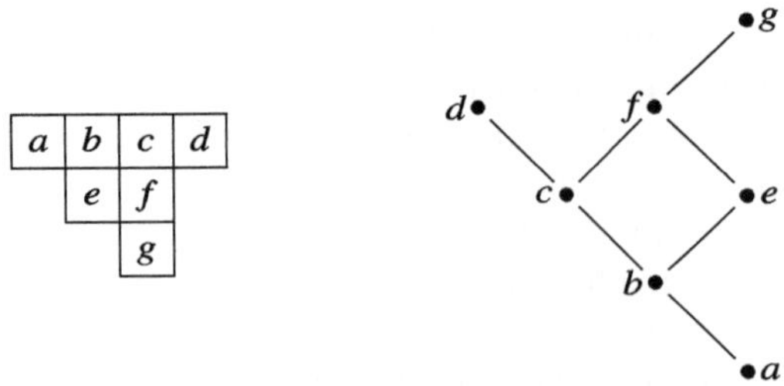

Fig. 1.5 The shifted diagram of shape (4, 2, 1) and its corresponding poset $P_{(4,2,1)}$

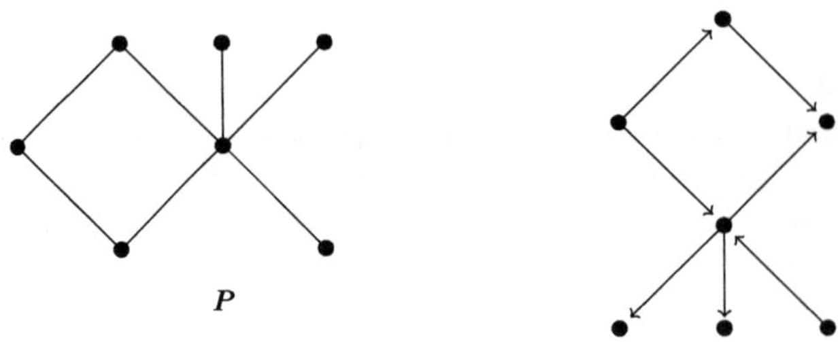

Fig. 1.6 A poset P and its corresponding directed acyclic graph

in graphs like we have in Hasse diagrams. To create relationships between vertices of a graph, we can make a directed graph by putting arrows on edges. For instance, in Fig. 1.6, we see an unlabeled poset P. For every covering relation $x \lessdot y$, we can make that edge directed by adding an arrow $x \to y$. With these arrows in place, the placement of the vertices on the page can be arbitrary. On the right in Fig. 1.6, we see the corresponding directed graph for the poset P. Notice that the vertices are in different positions on the page, but we can still tell what the covering relations are between elements.

By making a poset into a directed graph, we can begin to ask questions that are inspired by graph theory. The first we can ask: What cycles can we find on these graphs, if any? Consider a cycle on a directed graph. Let the elements in the cycle be $x_1, x_2, \ldots, x_k$. Therefore, on the directed graph, we have that $x_1 \to x_2 \to \cdots \to x_k \to x_1$. Since it is a cycle, we return to x_1 from x_k. Translating this back to what it means in a poset, we have that $x_1 < x_2 < \cdots < x_k < x_1$. So, if there is a cycle, then there is an element x_1 such that $x_1 < x_1$! This is not allowed in the definition of a poset, and so it must be the case that we have no cycles in the directed graph. When a graph has no cycles, we call it *acyclic*. This is, in fact, a characterization of the directed graphs arising from posets: All posets can be viewed as directed acyclic graphs, and vice versa.[1]

[1] Some authors require that the directed graph associated with a poset includes the directed edges for *all* relations in the poset, not just cover relations. In that case, you might call what we have created a "cover directed graph." If your directed graph includes all relations from the poset, it will still be acyclic.

When dealing with a directed graph, we call a vertex x a *source* if all edges incident to x are pointing away from x. We call a vertex y a *sink* if all edges incident to y are pointing into y. In terms of a poset P, a minimal element in P becomes a source in the directed acyclic graph and a maximal element in P become a sink in the directed acyclic graph.

Next you might wonder why we care about directed acyclic graphs. In a practical sense, directed acyclic graphs help us organize and solve all types of scheduling problems. They are easy for computers to store and manipulate. Let's take a look at one example of this type of problem and how it might be solved using graph theory.

Example 1.4 Kiera wants to plan her mathematics electives, and she is interested in the following nine courses. The prerequisites are given below. Find a sequence of courses that will allow Kiera to satisfy all prerequisites.

- MA215: (none)
- MA216: MA215
- MA222: (none)
- MA231: MA215
- MA232: MA216, MA231
- MA341: MA216, MA222
- MA369: MA232
- MA422: MA222, MA232
- MA426: MA231

First, let's draw a directed acyclic graph containing all the prerequisite information. See Fig. 1.7. We can find the required sequence by following an algorithm. See Chapter 12 in [8] for pseudo-code for this algorithm, which can be turned into computer code using the programming language of your choice.

1. Let $i = 1$.
2. Identify the in-degree of each vertex in the graph. Choose a vertex v_i with in-degree 0 as the next vertex in your sequence.
3. Remove v_i from your graph along with corresponding edges.

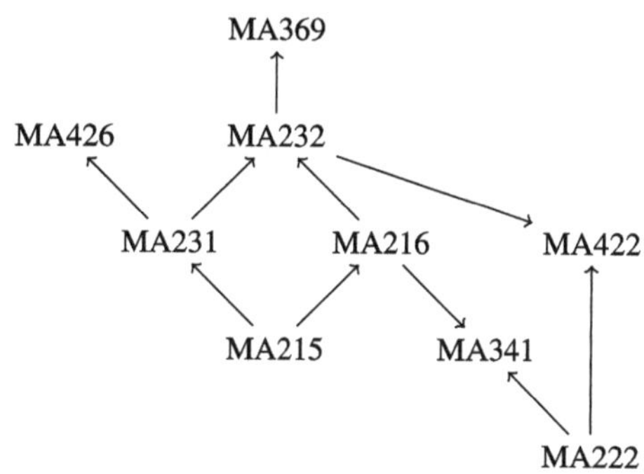

Fig. 1.7 The acyclic directed graph for Example 1.4

4. Increment i by 1, and repeat steps 2–4 until every vertex has been added to your sequence.

   For our example, one such sequence we get is

$$\text{MA215, MA216, MA231, MA426, MA232, MA369, MA222, MS341, MA422.}$$

□

There are many correct solutions to this scheduling problem. You might even wonder how many such sequences exist. In fact, we have unwittingly found a linear extension for this poset, even though we have not yet defined what that means. There are usually many linear extensions for a given poset. Let's explore those next.

1.3 Linear Extensions

So far, we have focused on sets with a *partial* order. You might be curious to know what kind of *total* order can be imposed on such a set. I'm so glad you asked! In order to discuss possible total orders on a poset, we first need to discuss labeled posets.

Definition 1.5 Let P and Q be posets. A map $\phi : P \to Q$ is *order preserving* if $x \leq_P y$ implies that $\phi(x) \leq_Q \phi(y)$. Similarly, a map $\sigma : P \to Q$ is *order reversing* if $x \leq_P y$ implies $\sigma(y) \leq_Q \sigma(x)$. An *isomorphism* is an order preserving bijection whose inverse is also order preserving. An *automorphism* is an isomorphism from P to itself, and an *anti-automorphism* is an order-reversing bijection from P to itself.

For any (finite) poset P, we can give it a *labeling*, which is a bijection $\omega : P \to \{1, 2, \ldots, n\}$, where P has n elements. If we want to emphasize the labeling for P, we refer to the poset as (P, ω). A labeling is *natural* if it is order preserving. See Fig. 1.8 for an example of a poset with a natural labeling. In general, we fix a natural labeling on P and let the elements of P correspond to $\{1, 2, \ldots, n\}$.

Once we have fixed a labeling for a given poset, we have a convenient naming system that allows us to form linear extensions. A *linear extension* is a total order

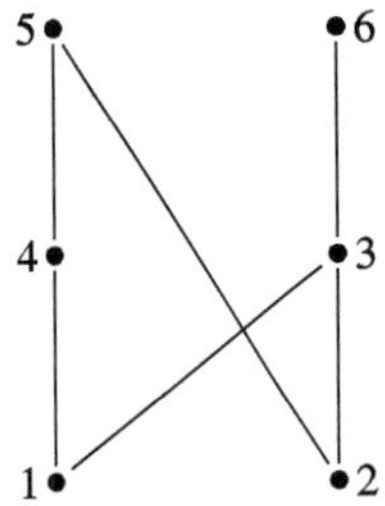

$E(P) =$
$\{123456,\ 123465,\ 123645,\ 124356,$
$124365,\ 124536,\ 142356,\ 142365,$
$142536,\ 213456,\ 213465,\ 213645,$
$214356,\ 214365,\ 214536\}.$

Fig. 1.8 A poset with a natural labeling and its set $E(P)$ of linear extensions

$x_1 \prec x_2 \prec \cdots \prec x_n$ on the elements of P such that $x_i \prec x_j$ if $x_i <_P x_j$. This means that if $x <_P y$, then x must appear to the left of y in the linear extension. We will only use the $\prec$ notation when we are discussing elements in a linear extension. Also, we can view a linear extension as a permutation $x_1 x_2 \cdots x_n$ in one-line notation, which means the permutation maps i to x_i. For further discussion of these definitions, see Example 1.5.

Let $E(P)$ be the set of linear extensions of P and $e(P)$ be the cardinality of $E(P)$. If $(P, \leq)$ is a poset and $x, y \in P$, let $P + xy$ denote the poset $(P, \leq')$, where $\leq'$ is the binary operation that extends $\leq$ by including the relations $x \leq' y$, $z \leq' y$ for all $z \leq x$, and $x \leq' w$ for all $y \leq w$. Therefore, $E(P + xy)$ is the set of linear extensions of P that have x before y and $e(P + xy)$ is the size of $E(P + xy)$. Notice that $e(P + xy) + e(P + yx) = e(P)$.

Example 1.5 Consider the poset in Fig. 1.8.[2] The labeling is natural, meaning it is an order preserving bijection $\omega : P \to \{1, 2, 3, 4, 5, 6\}$, since P has six elements. Also in the figure, we can see the list of 15 linear extensions of P. We can make a few observations. Since the poset dictates that $2 <_P 5$, then every linear extension must list 2 before 5 when reading from left to right. However, there is no relationship between 5 and 6 in the poset, and so there are linear extensions with $5 \prec 6$ and others with $6 \prec 5$.

In this example, $e(P) = 15$. Further,

$$\begin{aligned} &E(P + 56) \\ &\quad = \{123456, 124356, 124536, 142356, 142536, 213456, 214356, 214536\}, \end{aligned}$$

and so $e(P + 56) = 8$. □

Linear extensions of posets have been extensively studied, but there are still many unknown properties to explore. For instance, what can we say about the poset when looking at the list of its linear extensions? This is what we will discuss in the next section.

1.4 The 1/3–2/3 Conjecture

We are finally ready to learn about a famous conjecture in combinatorics: the $1/3-2/3$ Conjecture. We have covered all of the relevant machinery in the introduction; for this conjecture, we set up the following notation.

Let $(P, \leq)$ be a partially ordered set, and let n be the cardinality of P. Fix a natural labeling on P, so that the elements of P correspond to $\{1, 2, \ldots, n\}$. For distinct elements $x, y \in P$, we define $\mathbb{P}(x \prec y)$ to be the proportion of linear

[2] Figures 1.8, 1.9, 1.10, and 1.11 in this article first appeared in [12] as Figures 1.1, 1.2, 6.1, and 6.2, respectively.

extensions of P in which x comes before y. Notice that if P is a chain, then $\mathbb{P}(x \prec y) = 1$ whenever $x <_P y$. Since chains are so predictable, we exclude chains from the conjecture. We can also see that a poset with a $\hat{0}$ has $\mathbb{P}(\hat{0} \prec x) = 1$ for all $x > \hat{0}$, and so the proportion of linear extensions remains the same in $P - \{\hat{0}\}$. For this reason, we usually insist that posets have at least 2 minimal elements and, similarly, at least 2 maximal elements.

For $0 \leq \alpha \leq \frac{1}{2}$, we say (x, y) is an *α-balanced pair* if

$$\alpha \leq \mathbb{P}(x \prec y) \leq 1 - \alpha,$$

and that P is *α-balanced* if it has some α-balanced pair. Notice that if (x, y) is α-balanced, then (y, x) is α-balanced as well. We'll go through an example using this new terminology after stating the conjecture.

The $1/3 - 2/3$ **Conjecture**. Every finite partially ordered set that is not a chain has a $1/3$-balanced pair.

Example 1.6 Consider the poset in Fig. 1.8. This poset has 15 linear extensions total. If we consider just the elements 1 and 2, we can see that $\mathbb{P}(1 \prec 2) = \frac{9}{15}$. This is the proportion of linear extensions with 1 before 2. If we compare every pair of elements in P and organize our data into a matrix, we get

$$\begin{bmatrix} 0 & \mathbf{9} & 15 & 15 & 15 & 15 \\ \mathbf{6} & 0 & 15 & 12 & 15 & 15 \\ 0 & 0 & 0 & \mathbf{6} & 12 & 15 \\ 0 & 3 & \mathbf{9} & 0 & 15 & 13 \\ 0 & 0 & 3 & 0 & 0 & \mathbf{8} \\ 0 & 0 & 0 & 2 & \mathbf{7} & 0 \end{bmatrix}.$$

This matrix has for each entry (i, j) the number of linear extensions with i before j. For instance, P has 9 linear extensions with $1 \prec 2$ and 6 with $2 \prec 1$, and so the entry $(1, 2)$ in the matrix is 9, while the $(2, 1)$ entry is 6.

Notice that if $i <_P j$ in the poset, then the (i, j)th entry in the matrix is the number of linear extensions of P. The bold entries have values that are between $1/3$ and $2/3$ of the total number of linear extensions, which gives us the $1/3$-balanced pairs, namely, $(1, 2)$, $(2, 1)$, $(3, 4)$, $(4, 3)$, $(5, 6)$ and $(6, 5)$. □

We define the *balance constant* of P, denoted $\delta(P)$, to be

$$\delta(P) = \max_{x,y \in P} \min\{\mathbb{P}(x \prec y), \mathbb{P}(y \prec x)\}.$$

For any poset P not a chain, it must be that $0 < \delta(P) \leq 1/2$. If we consider the linear sum $P + Q$ of posets P and Q, we can see that all linear extensions of $P + Q$ are linear extensions of P concatenated with linear extensions of Q. Therefore, it is the case that $\delta(P + Q) = \max\{\delta(P), \delta(Q)\}$.

In the example in Fig. 1.8, P has a balance constant of $\frac{7}{15} \approx 0.4667$. So far, we have presented two equivalent ideas. The balance constant of a poset is greater than $1/3$ if and only if the poset has a $1/3$-balanced pair. We will use these two ideas interchangeably.

1.4.1 History

The conjecture was first proposed by Kislitsyn [10] in 1968, although many resources attribute the conjecture to having been formulated by Fredman [6] in 1976 and later independently by Linial [11] in 1984.

Given the age of the $1/3-2/3$ Conjecture, it should come as no surprise that partial results have been generated by many mathematicians. One of the first results appeared in 1984 by Linial [11], namely, that the $1/3-2/3$ Conjecture holds for posets of width 2, like in Example 1.6.

Theorem 1.1 ([11]) *Let $(P, \leq)$ be a poset of width exactly 2. Then, $\delta(P) \geq 1/3$.*

Aigner [1] showed further that posets of width exactly 2 fit into one of two categories: Either the poset is a linear sum of copies of the singleton poset and T (the poset in Fig. 1.9), or the poset has an α-balanced pair with $1/3 < \alpha < 2/3$. In fact, the only known posets that have a balance constant equal to exactly $1/3$ are the linear sums of singletons and T. As an exercise, you should check that T does have a balance constant of $1/3$. The poset of width 2 in Fig. 1.10 has a balance constant of $\frac{16}{45} \approx 0.3556$, and until recently, it was the poset with the smallest known balance constant greater than $1/3$ [3]. See Sect. 1.5.1 for my results regarding posets of width 2 that have smaller balance constants.

The smallest known balance constant for a poset with width strictly greater than 2 is $\frac{14}{39} \approx 0.3590$, as described in [3]. It belongs to the poset with 7 elements in Fig. 1.10. I searched all posets with up to 9 elements and found no posets with balance constant smaller that $\frac{14}{39}$ and width greater than 2.

There are many types of posets for which the conjecture has already been proven. Some of these concepts have not been covered in this chapter, but the interested reader is encouraged to find out more. The conjecture has been proved for posets of up to 11 elements [13], posets with height 2 [16], semiorders [2], posets with each element incomparable to at most 6 others [14], N-free posets [17], and posets whose

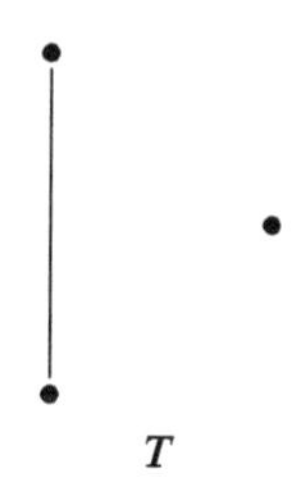

Fig. 1.9 The poset T with three elements and one relation

Fig. 1.10 Two posets with small balance constants

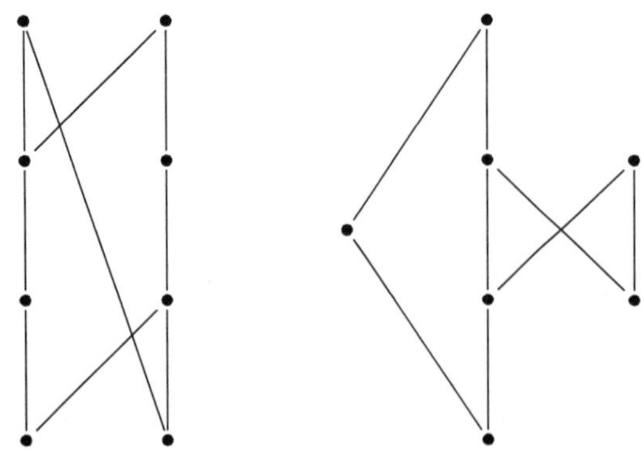

Hasse diagram is a tree [18]. While the proof of the 1/3 bound for a general poset remains elusive, in 1984, Kahn and Saks [9] proved that for any poset P, there is some pair $x, y \in P$ such that $\frac{3}{11} < \mathbb{P}(x \prec y) < \frac{8}{11}$. In 1995, Brightwell, Felsner, and Trotter [4] improved the bound to be $\frac{5-\sqrt{5}}{10} \leq \mathbb{P}(x \prec y) \leq \frac{5+\sqrt{5}}{10}$.[3] Refer to Brightwell's 1999 survey [3] on the conjecture for more information.

1.4.2 Contributions from Others

We now state a result of Zaguia, which will be useful to us in the later sections. The following definitions describe concepts introduced in [18], although here we refer to them with different names. Given x in a poset P, we let L_x and U_x denote the *strict lower* and *strict upper order ideals* generated by x, that is,

$$L_x = \{y \in P \mid y < x\} \quad \text{and} \quad U_x = \{y \in P \mid y > x\}.$$

Definition 1.6 Let P be a poset, and x and y be elements of P.

(a) We call the pair (x, y) *twin elements* if $L_x = L_y$ and $U_x = U_y$.
(b) We call the pair (x, y) *almost twin elements* if the following two conditions hold in P or in the dual of P:

  (i) $L_x = L_y$, and
  (ii) $U_x \setminus U_y$ and $U_y \setminus U_x$ are chains (possibly empty).

Figure 1.2 provides an example of a twin pair of elements, namely, the elements y and z in $P + Q$. In the same figure, elements b and c are an almost twin pair in the poset $P + Q$.

[3] For reference, $\frac{3}{11} \approx 0.2728$, $\frac{8}{11} \approx 0.7273$, $\frac{5-\sqrt{5}}{10} \approx 0.2764$, and $\frac{5+\sqrt{5}}{10} \approx 0.7236$.

We will see that a poset with twin elements is $1/2$-balanced in Sect. 1.5.2. In [18], Zaguia proves that a poset with an almost twin pair is $1/3$-balanced. In fact, he proves something stronger by generalizing the notion of an almost twin pair. He uses the term *good pair* to describe these elements, but we will call them *asymmetric*.

Definition 1.7 Let P be a poset. A pair (x, y) of elements of P is *asymmetric* if the following two conditions hold in P or in the dual of P:

(i) $L_x \subseteq L_y$ and $U_y \setminus U_x$ is a chain (possibly empty), and
(ii) $\mathbb{P}(x \prec y) \leq \frac{1}{2}$.

In fact, every almost twin pair of elements is also asymmetric, and Zaguia's proof focuses on asymmetric pairs. Since we refer to almost twin elements in Sect. 1.5.4, we include them in Theorem 1.2 for emphasis.

Theorem 1.2 ([18]) *A finite poset that has an asymmetric pair of elements or an almost twin pair of elements is* $1/3$*-balanced.*

It is important to keep in mind that many of the known results for the $1/3-2/3$ Conjecture are existence proofs and do not compute $\mathbb{P}(x \prec y)$ exactly for any pair (x, y) in the given poset. This is particularly true in the case of asymmetric pairs and almost twin pairs of elements. There are many examples of posets with asymmetric or almost twin pairs of elements that are not, in fact, the $1/3$-balanced pairs of the poset.

1.5 Contributions

1.5.1 Smallest Balance Constants

As previously mentioned, Aigner [1] proved that a poset of width 2 has a balance constant of $1/3$ if and only if it is the linear sum of the singleton poset and T from Fig. 1.9. One can then ask: Regardless of width, how close to $1/3$ can a balance constant be for a poset not of this form? We include the posets with the previously closest values in Fig. 1.10.

In [12], I present posets of width 2 that at the time had balance constants with the closest known values to $1/3$. In Fig. 1.11, poset A has $\delta(A) = \frac{6}{17} \approx 0.35294$. Poset B has $\delta(B) = \frac{60}{171} \approx 0.350877$. Poset C has $\delta(C) = \frac{37}{106} \approx 0.349057$.

I performed the analysis through code which inputs a poset, calculates its linear extensions, and finds $\mathbb{P}(x \prec y)$ for each pair $(x, y) \in P$. The code outputs the balance constant and can be set up to run for many posets at a time. Further research has already been done in this area. In [5], Chen found a family of posets with smaller balance constants, generalizing the poset C in Fig. 1.11. More recently, in [15], Sah improved the lower bound for the balance constant of a poset of width 2 to be $\dfrac{-3 + 5\sqrt{17}}{52} \approx 0.33876$.

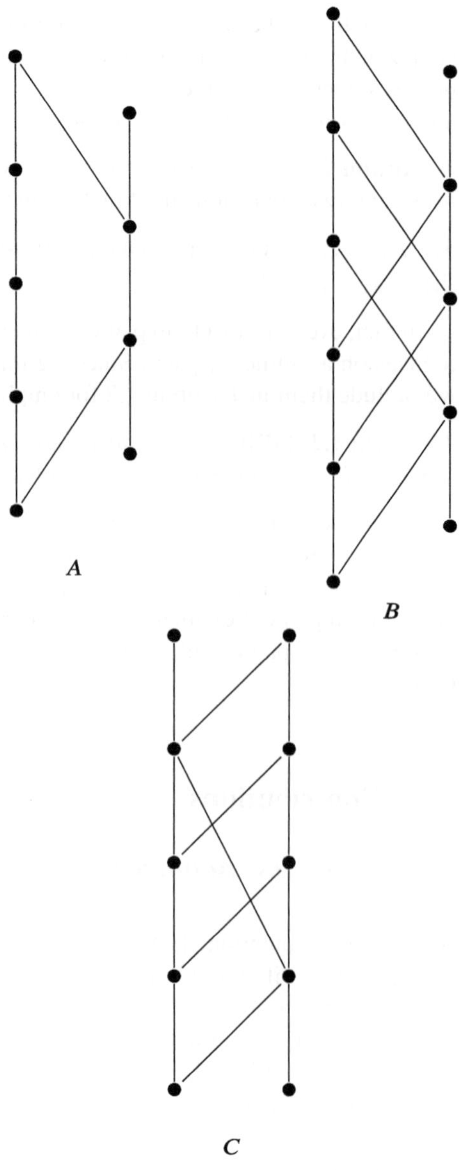

Fig. 1.11 Posets with the smallest balance constants greater than $1/3$

1.5.2 Automorphisms

We saw automorphisms of posets in Definition 1.5. It turns out a nontrivial automorphism of a poset is a useful way to identify the existence of $1/3$-balanced pairs. We begin with a result about linear extensions.

Proposition 1.1 ([12]) *A poset P with a nontrivial automorphism α has a nontrivial bijection on its linear extensions. Further,* $\mathbb{P}(x \prec y) = \mathbb{P}(\alpha(x) \prec \alpha(y))$ *for all* $x, y \in P$.

In [7], Ganter, Hafner, and Poguntke prove that a poset with a nontrivial automorphism will satisfy the $1/3 - 2/3$ Conjecture. Their short argument, which is worth checking out, is an excellent use of proof by contradiction.

Theorem 1.3 ([7]) *If a poset P has a nontrivial automorphism, then P is $1/3$-balanced.*

In our research, we became curious about not just $1/3$-balanced pairs but $1/2$-balanced pairs. There are many posets without $1/2$-balanced pairs, and we wanted to investigate when such a pair would exist. Next is a result describing one case when a poset has a balance constant of $1/2$.

Proposition 1.2 ([12]) *If a poset P has an automorphism with a cycle of length 2, then P is $1/2$-balanced. Further, if x and y are the elements in the cycle of length 2, then (x, y) is a $1/2$-balanced pair.*

Recall twin pairs of elements from Definition 1.6. Another example of a $1/2$-balanced poset is one with a twin pair of elements; in fact, the twin pair of elements is $1/2$-balanced in the poset. This fact is a corollary to Proposition 1.2.

Corollary 1.1 ([12]) *A poset P with a twin pair of elements is $1/2$-balanced.*

1.5.3 Lattices

One commonly studied type of poset is a lattice. Here, we discuss the general definition of a lattice before considering the conjecture for specific lattices in this section.

Definition 1.8 Let $S \subseteq P$ be a nonempty set of elements. A *lower bound* of S is an element z such that $z \leq u$ for all $u \in S$. The *greatest lower bound*, or *meet*, of S is a lower bound w of S such that if z is another lower bound of S, then $z \leq w$. If $S = \{x, y\}$, then the meet, if it exists, is denoted by $x \wedge y$.

Similarly, an *upper bound* of S is an element z such that $z \geq u$ for all $u \in S$. The *least upper bound*, or *join*, of S is an upper bound w of S such that if z is another upper bound of S, then $z \geq w$. If $S = \{x, y\}$, then the join, if it exists, is denoted by $x \vee y$.

Note that in general, a set $S \subseteq P$ could have zero, one, or many lower bounds or upper bounds. However, if a meet or join exists, it must be unique. We call a poset a *lattice* if every nonempty subset $S \subseteq P$ has a meet and a join.

Example 1.7 Consider the set of all subsets of $\{1, \ldots, n\}$, and define an order by $S \leq T$ when $S \subseteq T$ for $S, T \subseteq \{1, \ldots, n\}$. This is a lattice as it is closed under joins and meets, specifically, $S \vee T = S \cup T$ and $S \wedge T = S \cap T$. We call this poset

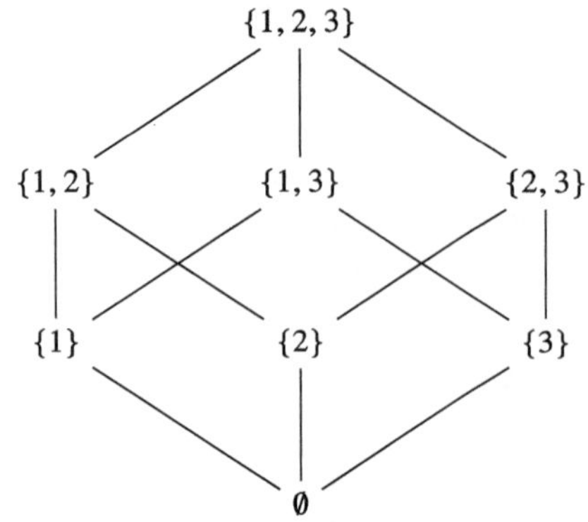

Fig. 1.12 The Boolean lattice B_3

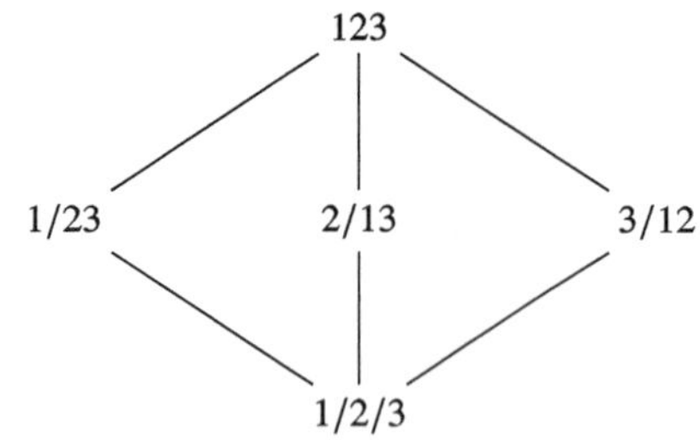

Fig. 1.13 The set partition lattice Π_3

the *Boolean lattice* of size n, denoted B_n. This poset has the empty set as its $\hat{0}$ and $\{1, \ldots, n\}$ as its $\hat{1}$. An example when $n = 3$ is given in Fig. 1.12.

In the case that $n = 1$, the poset B_1 is a chain, and so the $1/3 - 2/3$ Conjecture only applies to B_n when $n \geq 2$. The following is a corollary to Proposition 1.2. □

Corollary 1.2 ([12]) *For all $n \geq 2$, the Boolean lattice B_n has an automorphism with a cycle of length 2, and so the Boolean lattice is $1/2$-balanced.*

Example 1.8 Consider the set Π_n of all partitions of $\{1, \ldots, n\}$. A *partition*[4] is composed of pairwise disjoint, non-empty subsets $B_1, \ldots, B_k$ whose union is $\{1, \ldots, n\}$, and we write $B_1/\cdots/B_k$ for a partition of $\{1, \ldots, n\}$. Each B_i is called a *block* of the partition. For $\pi, \tau \in \Pi_n$, we say π is a *refinement* of τ if every block of π is contained in a block of τ. This produces a partial order on Π_n given by $\pi \leq \tau$ when π is a refinement of τ. This is a lattice as it is closed under joins and meets, specifically $\pi \vee \tau$ is the minimal partition that refines both π and τ and $\pi \wedge \tau$ is the maximal partition that is a refinement of both π and τ.

Hence, Π_n is called the *set partition lattice*. For brevity, we will write 12/3/4 for the set partition $\{\{1, 2\}, \{3\}, \{4\}\}$ and similarly for other partitions. The diagram of Π_3 is in Fig. 1.13. □

[4] This is different than an integer partition, which we saw in Sect. 1.2.2.

For $n = 1, 2$, Π_n is a chain, and so the $1/3 - 2/3$ Conjecture only applies to Π_n when $n \geq 3$. The following is a corollary to Proposition 1.2.

Corollary 1.3 ([12]) *For $n \geq 3$, the set partition lattice Π_n has an automorphism with a cycle of length 2, and so the set partition lattice is $1/2$-balanced.*

Example 1.9 Let $q = p^m$ be some power of a prime p, and consider the finite vector space $V = \mathbb{F}_q^n$, which is the set of n-tuples of elements from the finite field $\mathbb{F}_q$. The *subspace lattice* $L_n(q)$ consists of the set of all subspaces of $\mathbb{F}_q^n$ ordered by inclusion. We can see that for two subspaces $W, W' \subseteq \mathbb{F}_q^n$, their meet $W \wedge W'$ is $W \cap W'$ and their join $W \vee W'$ is $W + W'$. As $L_n(q)$ is closed under meets and joins, $L_n(q)$ is a lattice. □

For $n = 1$, $L_n(q)$ is a chain, and so the $1/3 - 2/3$ Conjecture only applies to $L_n(q)$ when $n \geq 2$. The following is a corollary to Proposition 1.2.

Corollary 1.4 ([12]) *For $n \geq 2$, the subspace lattice $L_n(q)$ has an automorphism with a cycle of length 2, and so the subspace lattice is $1/2$-balanced.*

1.5.4 Posets from Young Diagrams

Recall that in Sect. 1.2.2, we saw the definition of Young diagrams as combinatorial objects. We also saw in Fig. 1.5 how left-justified, shifted, and skew diagrams can be used to create posets. One special case of poset we can obtain is from the partition $(\lambda_1^{m_1})$, which is m_1 rows of cells each of length λ_1. In this case, this poset is not only $1/3$-balanced, but the proof specifically finds the $1/3$-balanced pair. This is unique, since many results toward the conjecture are existence proofs.

Theorem 1.4 ([12]) *Consider the integer partition $(\lambda_1^{m_1})$ and its corresponding poset $P_{(\lambda_1^{m_1})}$. Then, $P_{(\lambda_1^{m_1})}$ has a $1/3$-balanced pair.*

As we also saw in Sect. 1.2.2, the cells of a Young diagram can be filled with integers from 1 to n using each number exactly once and increasing in the rows and columns to form a standard Young tableau. In a similar manner, we define *shifted standard Young tableaux* and *skew standard Young tableaux*. For a given standard Young tableau, we can see that each cell, or element of the poset, has been assigned a number. Further, we have that if $x \leq_P y$, then the number assigned to x is smaller than the number assigned to y. This means we have a linear extension of the poset. In fact, the standard (shifted) (skew) Young tableaux for a given diagram are in bijective correspondence with linear extensions of the poset. Therefore, in the results regarding linear extensions for these posets, we rely on methods used for standard (shifted) (skew) Young tableaux instead.

We next present a generalized version of Theorem 1.4. For any type of diagram, we use the notation (i, j) to refer to the cell in the ith row and jth column. For instance, the $(2, 2)$ cell in the shifted diagram in Fig. 1.5 is labeled with an e.

Theorem 1.5 ([12]) *Let $P_{\lambda/\mu}$ be the poset corresponding to the diagram λ/μ, where μ could be an empty partition, and assume $P_{\lambda/\mu}$ is not a chain. If λ/μ is a Young diagram, shifted diagram, or skew diagram, $P_{\lambda/\mu}$ is a 1/3-balanced poset.*

The proof of this theorem finds almost twin pairs of elements in each type of diagram and relies on Zaguia's result in [18] for the existence of the 1/3-balanced pair.

1.6 Future Work

In Sect. 1.5.1, we discussed finding posets of width 2 with small balance constants. Future work in this direction would continue to search for posets with balance constants approaching 1/3, and it would be ideal to find a pattern among these posets.

In Sect. 1.5.2, we discussed one special case when posets are 1/2-balanced. Being 1/2-balanced is a much stronger requirement than being 1/3-balanced. It would be interesting to find more types of posets that are 1/2-balanced or characterize all such posets.

In Sect. 1.5.4, we discussed the poset $P_{(\lambda_1^{m_1})}$ in Theorem 1.4. One can observe that this particular type of poset is also the product of chains $C_{\lambda_1} \times C_{m_1}$. We were motivated to study products of chains as they are isomorphic to divisor lattices. A *divisor lattice* is the set of divisors of n partially ordered by $x \leq y$ if x divides y. It is a lattice as the meet of two elements is their greatest common divisor and the join of two elements is their least common multiple. By further exploring products of chains, we hope to prove the conjecture for divisor lattices, and so one future direction this work could take is to prove that a product of k chains is 1/3-balanced, for $k \geq 3$.

Speaking on the 1/3−2/3 Conjecture at conferences has provided me with quality feedback and general excitement in the combinatorics community for future work. There are many other directions of research this question can take, as evidenced by the number of suggestions received after presenting. I encourage others to consider working on this question as well.

References

1. Aigner, M.: A note on merging. Order **2**(3), 257–264 (1985)
2. Brightwell, G.R.: Semiorders and the 1/3–2/3 conjecture. Order **5**(4), 369–380 (1989)
3. Brightwell, G.: Balanced pairs in partial orders. Discrete Math. **201**(1–3), 25–52 (1999)
4. Brightwell, G.R., Felsner, S., Trotter, W.T.: Balancing pairs and the cross product conjecture. Order **12**(4), 327–349 (1995)
5. Chen, E.: A family of partially ordered sets with small balance constant. Electron. J. Comb. **25**(4), P4.43 (2018)

6. Fredman, M.L.: How good is the information theory bound in sorting? Theor. Comput. Sci. **1**(4), 355–361 (1976)
7. Ganter, B., Häfner, G., Poguntke, W.: On linear extensions of ordered sets with a symmetry. Discrete Math. **63**(2–3), 153–156 (1987)
8. Goodrich, M.T., Tamassia, R., Mount, D.M.: Data Structures and Algorithms in C++. Wiley, Hoboken (2004)
9. Kahn, J., Saks, M.: Balancing poset extensions. Order **1**(2), 113–126 (1984)
10. Kislitsyn, S.: Finite partially ordered sets and their associated sets of permutations. Matematicheskiye Zametki **4**(5), 511–518 (1968)
11. Linial, N.: The information-theoretic bound is good for merging. SIAM J. Comput. **13**(4), 795–801 (1984)
12. Olson, E.J., Sagan, B.E.: On the 1/3–2/3 conjecture. Order **35**, 581–596 (2018)
13. Peczarski, M.: The gold partition conjecture. Order **23**(1), 89–95 (2006)
14. Peczarski, M.: The gold partition conjecture for 6-thin posets. Order **25**(2), 91–103 (2008)
15. Sah, A.: Improving the 1/3–2/3 conjecture for width two posets. Combinatorica **41**, 99–126 (2021)
16. Trotter, W., Gehrlein, W., Fishburn, P.: Balance theorems for height-2 posets. Order **9**(1), 43–53 (1992)
17. Zaguia, I.: The 1/3-2/3 conjecture for N-free ordered sets. Electron. J. Comb. **19**(2), 1–5 (2012)
18. Zaguia, I.: The 1/3-2/3 conjecture for ordered sets whose cover graph is a forest. Order **36**, 335–347 (2019)

Chapter 2
Group Actions and Riemann Surfaces

Jennifer Paulhus

Abstract Riemann surfaces are fascinating topological objects with extra structure which allows us to translate ideas from analysis onto them. Additionally, groups can describe ways to permute the elements of a Riemann surface through a process called a group action. To give you a sense of what my field of research entails, we first need to understand group actions and Riemann surfaces. While this may seem like a tall order, you can grasp the ideas of these topics with just a first semester undergraduate course in both algebra and analysis, along with a basic knowledge of the complex numbers (such as roots of unity). An undergraduate course in topology may aid your understanding, but I define key ideas along the way.

Keywords Group actions · Riemann surfaces · Automorphisms of Riemann surfaces · Topology · Group theory · Complex analysis · Jacobian varieties · Mathematical databases · Curves · Riemann Existence Theorem

2.1 Biography/Career Trajectory

My mother likes to tell a story about me from a month or so into kindergarten. I came home one afternoon and she asked me how school had been. I responded that it was awful because I did not learn any new math. No one in my family worked in any field related to mathematics, so this response was quite unexpected. My parents, perhaps sensing a prophetic statement, started to encourage my mathematical thinking. In my office today, I have a Christmas present they bought me: a (still) functioning *Speak & Math*, part of Texas Instruments' triumvirate of early 1980's electronic educational toys (including the *Speak & Spell* and *Speak & Read*). I also crushed babysitters at the Milton Bradley *Memory* card matching game.

J. Paulhus (✉)
Grinnell College, Grinnell, IA, USA
e-mail: jpaulhus@mtholyoke.edu

A. Brisbin et al. (eds.), *Research Connections*, Association for Women in Mathematics Series 36, https://doi.org/10.1007/978-3-031-81217-0_2

Middle school and high school passed by in a blur. For a period of time, I planned to become an egyptologist, but an experience digging at an archeological site in the hot sun disabused me of that notion. I also spent some time wanting to become a commercial pilot, and I remember having the vague idea that being an engineer sounded fun. Meanwhile, the math I learned came very naturally, I participated in some math competitions, and I had no problem remembering just about everything I learned in my math classes.

As high school wound down, people told me I should go to college and major in math. But I am a stubborn person, and the best way to get me not to do something is to tell me I should do it. I went off to college planning to major in economics or history or philosophy, prepare for a career in the foreign service, and one day save the world. The fatal flaw in my plan started to show almost immediately. My economics course was dull, and I really loved my math classes. And so I found myself declaring a math major my second year of college. My math professors were kind and approachable, and I really enjoyed learning about proofs.

Somewhere in the middle of college, math started to get hard for me. Today, we have words like "growth mindset" and "fixed mindset," and with 25 years of hindsight, I can see that at the time, I had a strong fixed mindset. I assumed natural talent was the main driver for success in math, and I did not recognize the role of hard work and deep thinking. When math got hard, I did not understand why I could not immediately figure out all the material, I did not know how to grapple with complicated concepts, and I could not figure out how to fix any of that. My fixed mindset started to inhibit my progress during spring semester of my second year of college. I could pattern match well so my grades were still good, but I had the unsettling notion that I just did not have a strong grasp of the material. That summer I participated in the Carleton Summer Mathematics Program. I really enjoyed the classes, and the other students were great, but these unresolved issues continued to cause me anxiety, and I left the program early.

My junior year of college was better in some ways. I took my first abstract algebra class, and enjoyed the topics, but much of my success in that class was due to a natural affinity for the content. By the end of the year, I still did not feel comfortable with large parts of the material I had supposedly learned. After my junior year, I participated in a National Science Foundation Research Experience for Undergraduates (REU) at Rose-Hulman Institute of Technology. Overall, the summer was successful and gave me a sense that advanced research might be something I would enjoy. But there was still a niggling worry that I was getting by with pattern matching, and a surface level understanding.

During my senior year of college, I applied to Ph.D. programs in math, and also the Jesuit Volunteer Corps (a domestic Peace Corps of sorts). You can see that I was still attempting to save the world! I got five rejection letters from graduate programs, before I got accepted to the University of Illinois at Urbana-Champaign (UIUC) and one other university. I visited UIUC and enjoyed conversations with faculty and graduate students. The department also was strong in research areas I thought I wanted to go into. Meanwhile, I was getting advice to go immediately to graduate school, and that if I took time off to do the Jesuit Volunteer Corps first, I would forget too much math by the time I returned to school. So I chose graduate school.

In the weeks leading up to my move to Urbana-Champaign, I had a recurring dream that I got to UIUC and was told by one of the support staff that there had been a mistake and I had not actually been accepted. Boy was that an inauspicious way to start graduate school. My first year was rough. I was burned out from my undergraduate years, still did not really understand how to think deeply about math, and remained stuck in a fixed mindset. I spent the bulk of my mental energy worrying that I was never going to be able to understand the material. By the beginning of the spring semester, it was clear I needed a break.

At the time, Americorps had a program in Urbana-Champaign which I was accepted to, and I took a leave of absence from graduate school. I worked in a homeless shelter, appreciated having a job that I could leave behind at the end of each day, realized I do not have what it takes to be a social worker, and even started to miss math.

I returned to graduate school the next year with a little distance and perspective, and the beginning of an understanding that mathematics is hard work for everyone. If I had the whole thing to do over, I would have taken that year off before I went to graduate school. It did take me a little while to get caught back up after the year away, but my attitude and persistence were much improved. Graduate school progressed more or less from there. Looking back on it, I cannot quite determine what turned the ship around. It was still difficult, but small successes like passing the first couple qualifying exams begat more confidence which in turn begat more success. A certain amount of stubbornness on my part helped too.

My husband and I met at the Rose-Hulman REU. (We like to tell people outside of academics that we met at "math camp." It helps our nerd cred.) We decided early in our relationship that our individual careers were vitally important to each of us, and this informed how we made choices when solving our two-body problem. We did go to graduate school together, but UIUC was one of only two universities we both applied to, and it had strong programs for both of our mathematical interests. As graduate school ended, our plan was to each take the strongest offer we got. We hoped to trade time together early in our careers in favor of jobs we both liked together in the long run, and we felt that the stronger we were individually, the greater the likelihood we would both find enjoyable jobs in one location eventually. So we spent the better part of 4 years at jobs in different parts of the country. I accepted a postdoc at Kansas State University, while he went to Dartmouth College. If you think getting from the East Coast to the West Coast is hard, try getting from Manhattan, Kansas to Hanover, New Hampshire!

By the end of my third year in Kansas, he was settling into a tenure track position at Grinnell College, but they did not immediately have a job for me. So I accepted a tenure track job at Villanova University, and we both applied for jobs again the next year. By then, Grinnell College did have a job opening, and in my general research area. There was a certain amount of serendipity to the whole thing. We were able to find jobs we both like in a tiny town at a small college in Iowa, not in the Philadelphia area where Villanova is (and many other colleges and universities are).

My advice for solving the two-body problem is to be patient, persistent, and flexible. I still do recommend taking the best jobs you each get out of graduate school, if your relationship can handle it. Most of the couples I know who took jobs apart in the short term found something they were both happy with in the long run. This advice, of course, does not work for everyone. Regardless, patience, persistence, and a bit of flexibility often lead to success in the end.

But enough about me. Let's talk about the mathematics I think about ...

2.2 Introduction

Perhaps the most ubiquitous application of groups in modern mathematics is via group actions. We define group actions formally in the next section, but first I want to give you an idea of the topic. Given a set of objects, a group *acts on* the set if each element of the group can represent some shuffling of the set. Applications of this topic range from counting colorings up to some equivalence in combinatorics, to proving classical congruences in number theory, to studying automorphisms of surfaces. We will talk about the last topic later. Even physicists and chemists get in on the action (pun intended!) by, for example, interpreting problems about molecular structures or systems with rotational symmetries as problems of groups acting on vector spaces.

Here is a geometric example. Take your favorite regular polygon (regular means all sides are the same length and all angles are the same). My favorite is the regular pentagon. Place the shape on paper with one point (also called a *vertex*) at the top (north), label the vertices (five in my case), and mark where the vertices are located on the paper. Imagine lifting the shape up, moving it around without stretching or shrinking it, and then putting it back down to cover the same outline on the paper. In mathematical terms, the valid rearrangements of the shape on paper are called *symmetries*. Pause for a minute and find some symmetries of your favorite polygon.

Check out Fig. 2.1[1] for some examples. One symmetry is rotating the shape clockwise, so that each vertex moves to the one to the right of it. We could of course rotate again, and again, and again, but once we rotate five times, the object is back to its original positioning. Notice that rotating once counterclockwise has the same outcome as rotating clockwise 4 times. In Fig. 2.1, we also show reflection across the line from the vertex at the top to the middle of the bottom edge (or from the vertex at the top to the vertex at the bottom if your favorite regular polygon has an even number of vertices).

To create more symmetries, we can compose two known symmetries, thus forming a group with operation composition. You may recognize this group from abstract algebra: the *dihedral group*. There is some disagreement on how to denote this group. I call the group of symmetries on the regular n-gon D_n with $|D_n| = 2n$

[1] The author thanks Joseph Mileti for creating all the figures in this article.

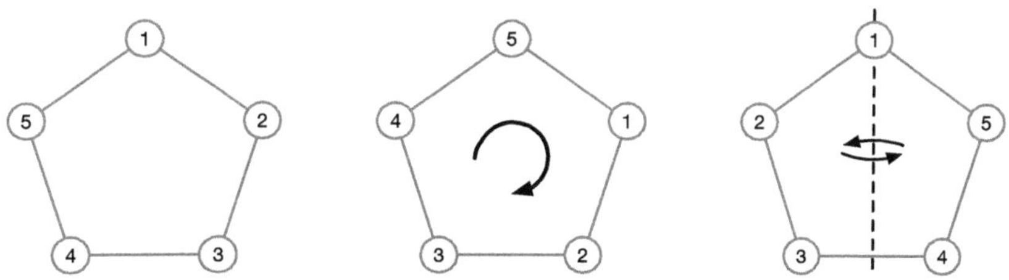

Fig. 2.1 From left to right: the regular pentagon as we describe positioned on paper, rotation by 72°, and reflection

but be warned that some books and research papers call the same group D_{2n}. This is what passes as drama in mathematics! Each symmetry (that is each element of the dihedral group) describes a way to shuffle the vertices of an n-gon, and so we come to our first example of a group action: D_n acts on the set of vertices of an n-gon.

In your linear algebra class you were introduced to linear transformations and a handy way of representing them: matrices. What are linear transformations? Functions from vectors to vectors (with some important additional properties, i.e., they preserve addition and scalar multiplication). In the case of $\mathbb{R}^2$, what does applying a linear transformation do? It moves vectors around the Cartesian plane. As an example, take the matrix $-I = \begin{bmatrix} -1 & 0 \\ 0 & -1 \end{bmatrix}$. It sends each vector to the corresponding vector in the opposite direction, rotating all vectors by 180°, except the zero vector which it fixes. What about a matrix like $M = \begin{bmatrix} 2 & 0 \\ 0 & 3 \end{bmatrix}$? It stretches a vector by a multiple of 2 in the x direction and by a multiple of 3 in the y direction, so each vector will get sent to one that has twice the x component and three times the y component, again fixing the zero vector.

Consider the linear transformation represented by $M = \begin{bmatrix} 4 & -3 \\ -1 & 2 \end{bmatrix}$. Think for a moment about what happens to vectors along the line $y = x$. They are fixed, precisely because $\begin{bmatrix} 1 \\ 1 \end{bmatrix}$ is an eigenvector with eigenvalue 1. What is the other eigenvalue and what does it tell you about how the linear transformation moves the corresponding eigenvectors? To describe what M does to the remaining (non-eigenvector) vectors, choose a basis of $\mathbb{R}^2$ consisting of one eigenvector for each of the two eigenvalues, and rewrite each vector in terms of that basis. The linear transformation will stretch the original vector by the amount of the component corresponding to the first eigenvector and stretch it by five times the amount of the component corresponding to the second eigenvector. (This should give you a hint for what the other eigenvalue is!)

And so we have our second example of a group action. The set of all 2×2 matrices with entries in $\mathbb{R}$ describes a way to shuffle the vectors in $\mathbb{R}^2$. You might be a bit worried, because the set of all 2×2 matrices with entries in $\mathbb{R}$ does not form a group under the operation of matrix multiplication (which axiom(s) fail?). We will address your concerns in the next section.

My goal in this paper is to introduce you to special group actions which I use in my own research. The sets are topological objects called *Riemann surfaces* (named for German mathematician Bernhard Riemann who first defined them), and there are many different groups which can act on these sets. In order to understand this topic, we need to first understand group actions (Sect. 2.3) and Riemann surfaces (Sect. 2.4). In Sect. 2.5, I describe what it means for groups to act on Riemann surfaces and then connect this to my own research (current and future) in Sects. 2.6 and 2.7. In an attempt to keep the exposition from getting cluttered, precise definitions not essential to a first understanding of the subject are in their own Sect. 2.8. Throughout the paper, I prompt you to pause and consider some questions. If this is your first time grappling with these topics, I highly recommend you stop reading at these points and think for a bit (probably with paper and pencil).

Finally, here is some standard notation I use: G is a group and 1_G its identity element, $|G|$ is the cardinality of G (size as a set), and $o(g)$ is the order of an element $g \in G$. For a complex number $z = x + yi$, $\text{Re}(z)$ is x, the real part of z, and $\text{Im}(z)$ is y, the imaginary part of z. Also $|z|$ is the *modulus* of z, defined as $|z| = \sqrt{x^2 + y^2}$.

2.3 Group Actions

The basic idea of a group action, as we saw in the introduction, is that elements of a group can provide information about how to "move around" (permute) the elements of a set, possibly keeping some elements fixed. The dihedral group elements give information about how to move around the vertices of a regular polygon. The group of 2×2 invertible matrices with entries in $\mathbb{R}$ describe a way to send vectors in $\mathbb{R}^2$ to each other. Notice, since we have restricted to invertible matrices, we now have a group acting on a set!

How do we formalize the idea that each element of the group permutes the set S? We define a bijective function $\varphi_g : S \to S$ for each $g \in G$ which describe how each group element moves around the elements of S. As with algebraic objects such as groups or rings, group actions have axioms. It is not enough that an element of the group merely shuffles elements of the set. There should be a nod to the group structure; otherwise, it would just be a set acting bijectively on a set! For example, we should expect the identity element of the group to behave as the identity element always does and leave all the elements of the set alone. Also, multiplication in the group should correspond to "multiplying" the functions φ_g defining the actions, except the natural way to "multiply" functions is to compose them. If we multiply two elements of the group together first and then consider what their product does to the set, that should be the same shuffling of the set as applying the second group element (numbered from left to right) to the set and then applying the first group element to the once shuffled set. In precise mathematical statements:

Definition 2.1 Let G be a group and S a set. A **group action** is a set of functions $\varphi_g : S \to S$, one function for each element $g \in G$, so that φ_{1_G} is the identity map, and $\varphi_{g_1 g_2} = \varphi_{g_1} \circ \varphi_{g_2}$.

We have seen two examples of group actions in the introduction. How do these examples satisfy the definition?

Example 2.1 When S is the set of vertices of a regular n-gon and $G = D_n$, then for each $g \in G$, the map φ_g is defined by sending each vertex to the vertex it goes to under the symmetry g. Note φ_{1_G} fixes all vertices, and since the group operation of G is function composition, the second group action axiom also follows. □

Example 2.2 The set of all 2×2 matrices with entries in $\mathbb{R}$ under the operation of matrix multiplication is not a group as not all matrices have multiplicative inverses, so we must restrict ourselves to matrices with nonzero determinant. With that restriction, and with S the set of vectors in $\mathbb{R}^2$, the discussion in the introduction naturally defines the group action. For each invertible 2×2 matrix A, the map $\varphi_A : S \to S$ is defined by multiplying the matrix A by each vector in $\mathbb{R}^2$. It is clear that the identity element of the group (the matrix I) fixes all vectors. The second group action axiom is satisfied if $(A_1 A_2)(v) = A_1(A_2(v))$ which is true simply by the way we define matrix multiplication. □

Example 2.3 A group can act on itself. Here the set S is equal to the group G. One way to describe an action of G on itself is by defining $\varphi_g : S \to S$ as $\varphi_g(h) = ghg^{-1}$ (*conjugation* by g). Pause here to check that the group action axioms are satisfied. Can you also come up with a different action of a group on itself? □

You may be worried that in the definition, I did not specify that φ_g must be a bijection. Fear not! The map φ_g has an inverse for each $g \in G$: the function $\varphi_{g^{-1}}$. Use the definition to show that this is, indeed, the inverse. You might see other definitions if you search for the phrase "group action" on the internet or in algebra books. For example, given a group G and a set S, a group action is a map from $G \times S$ to S with the image of the value (g, s) called $g \cdot s$, and with two properties: (1) $1_G \cdot s = s$ for all $s \in S$ and (2) $(gh) \cdot s = g \cdot (h \cdot s)$ for all $g, h \in G$ and $s \in S$. Although the group action axioms using these other definitions may look different to you on first inspection, it is a nice exercise to prove they are actually equivalent to the definition I give above.

Now that we have a formal definition and a couple of examples under our belt, there are a few sets we need to define which will be important later in the paper, and which allow us to more richly describe each particular group action. Elements of the set S that are fixed (unmoved) by an element of the group play an important role in understanding group actions. If we pick one element of the set S, call it s, then the set of elements of G which fix s is the **stabilizer of** s. In math symbols, we write $G_s = \{g \in G \mid \varphi_g(s) = s\}$. To get comfortable with this definition, here are some questions to consider. Where does the stabilizer live, i.e., what is it a subset of? Can the stabilizer ever be the empty set? Does the stabilizer have any additional structure to it, e.g., is it a subgroup?

Example 2.4 For the action of D_n on vertices of a regular n-gon, reflection is in the stabilizer of the first vertex, since this symmetry fixes that vertex, and if n is even, it is also in the stabilizer of the $\left(\frac{n}{2}+1\right)$st vertex, since the line we reflect across also goes through this vertex. For the action of linear transformations on $\mathbb{R}^2$, the stabilizer of $\begin{bmatrix}1\\1\end{bmatrix}$ includes the matrix $M = \begin{bmatrix}4 & -3\\-1 & 2\end{bmatrix}$. For an abelian group acting on itself by conjugation, the stabilizer of every element is the whole group, since $ghg^{-1} = gg^{-1}h = h$ for all $g, h \in G$. □

Connected to the stabilizer are two other adjectives used to describe a group action.

Definition 2.2 A group action is **faithful** if for all nonidentity elements $g \in G$, there is an $s \in S$ so that $\varphi_g(s) \neq s$. This definition means that no nonidentity element of G can be in all the stabilizers or that the intersection of all stabilizers of elements in S contains only the identity element. Here is a good place to pause and make sure you believe the first and second sentence are really saying the same thing mathematically.

Definition 2.3 A group action is **free** if whenever there is some $s \in S$ so that $\varphi_g(s) = s$, then g is the identity. Any such g would be in the stabilizer of s, and so this definition says that the stabilizers of all elements of S are trivial (only contain the identity element). A free action is also faithful, although the converse is not generally true.

Another important set is the **orbit** of an $s \in S$, denoted O_s. This is the set of elements of S which can be "reached" from s under the action. Alternatively, O_s is the image of s under all of the functions φ_g. In math symbols, we write $O_s = \{t \in S \mid \varphi_g(s) = t \text{ for some } g \in G\}$. For orbits, consider the same questions I asked you about the stabilizer.

Example 2.5 When D_n acts on the set of vertices of a regular n-gon, there is only one orbit, as repeated applications of rotation will move any vertex to any other vertex. In the case of matrices acting on vectors in $\mathbb{R}^2$, there are two orbits. One contains just the zero vector, and all the other vectors are in the other orbit. To show this, prove that given any two nonzero vectors $\mathbf{v_1}$ and $\mathbf{v_2}$ in $\mathbb{R}^2$, there is an invertible matrix M so that $M\mathbf{v_1} = \mathbf{v_2}$. For a group acting on itself by conjugation, the orbits are given the name *conjugacy classes*. □

When a group G acts on a set S, we can create the set of all orbits of this action. It turns out that if we take $r, s \in S$, then O_r and O_s either intersect trivially or are identical (in math terminology we say that the orbits *partition* the set S). One way to prove this is to show that the relation defined as

$$r \sim s \text{ if and only if there is a } g \in G \text{ so that } \varphi_g(r) = s$$

is an equivalence relation, because an equivalence relation partitions a set. Proving the assertions above would be a good refresher on equivalence relations.

The terminology of group actions is naturally descriptive. The *stabilizer* is the set of group elements which keep an element of the set stable, or unmoving! Think of *orbit* like in the phrase "my social orbit" in reference to your extended circle of friends and social acquaintances.

2.4 Riemann Surfaces

One of the key concepts in a calculus of one variable course is that "nice" functions (usually meaning functions that are differentiable everywhere) are those that look like lines close up. Lines are basic objects we all understand well, so if we can approximate a complicated function with a bunch of lines, we reduce the complexity of the function, at least locally. We say *locally* to mean in a small neighborhood of a point. But, we also want those different lines to interact well with each other. The step function can be approximated by lines, but those discontinuous points (the jumps) make the step function a less desirable function, from a calculus point of view. We cannot, for instance, define a derivative everywhere on this function. Or consider the absolute value function. It can be approximated by two lines ($y = x$ if $x \geq 0$ and $y = -x$ if $x \leq 0$), but the slope of those two lines do not coincide on their intersection (at $x = 0$), and so we cannot define a derivative at the point $x = 0$. No matter how tiny the neighborhood around the point $x = 0$ is, the absolute value in that neighborhood will never look like a line, it will always be V-shaped. See Fig. 2.2 if you forget what these functions look like.

A similar concept exists in higher dimensions. Instead of functions in one variable, we consider *topological spaces*. If you have not seen any topology, the definition of topological spaces is in Sect. 2.8. Following convention, we sometimes call topological spaces simply "spaces." We will talk of *points* on spaces, and a *neighborhood* around the point, by which we mean an open set containing the point. Informally, the topological spaces called *Riemann surfaces* we discuss in this paper are like surfaces in $\mathbb{R}^3$, specifically those surfaces which locally resemble a plane.

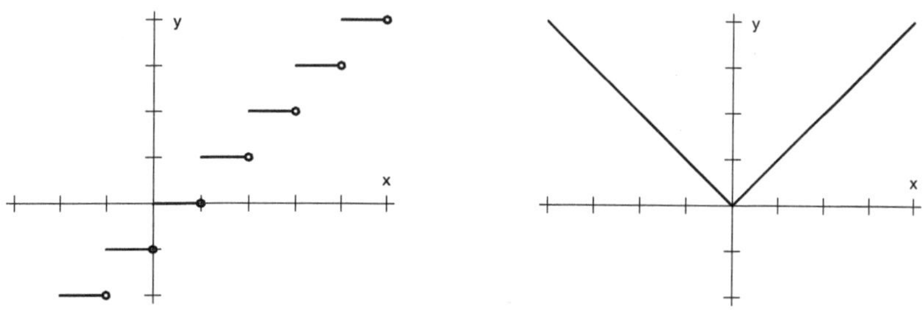

Fig. 2.2 The step function $f(x) = \lfloor x \rfloor$ (left) and the absolute value function $f(x) = |x|$ (right)

While most of the analysis you have seen so far likely featured real numbers, there are several advantages to analysis over the complex numbers. In real analysis, there are functions that are differentiable but whose derivatives are not differentiable. A way to construct such a function is to start with your favorite continuous, but not differentiable, function. Mine is $f(x) = |x|$. Then the Fundamental Theorem of Calculus produces a function which is differentiable, but whose derivative is your favorite continuous-but-not-differentiable function. In my case, one example would be the function $F(x) = \int_{-1}^{t} |t|\,dt$ on the interval $[-1, 1]$. In real analysis, there are also functions that do not equal their Taylor series. The function $g(x) = \begin{cases} e^{-1/x^2} & \text{if } x \neq 0 \\ 0 & \text{if } x = 0 \end{cases}$ is an example. It takes some work with limits, but you can show all of the derivatives of $g(x)$ at $x = 0$ evaluate to 0. So the Taylor series expansion at 0 is just the zero function, which is certainly not equal to $g(x)$ anywhere except at $x = 0$.

Neither of these scenarios occur in complex analysis: If a function is differentiable, it is infinitely differentiable, and the Taylor series of a differentiable complex function will equal the function inside the radius of convergence (of the Taylor series). Therefore, we want to study complicated topological spaces which look like the complex plane locally.

Here is how we formalize this idea on such spaces. Around any point on the space, we define a map from a neighborhood of that point to some open subset of the complex plane. If we want to apply the results of complex analysis to the space, we use the map to move from the space to the complex plane, apply the analysis results there, and then use the inverse of the map to return to the space. This plays an analogous role to that of describing complicated real single variable functions locally by lines. In that case, think of the map as sending a neighborhood near a point p to a line (the *tangent line*, in fact).

There are a few technicalities to work out. We need the map to have an inverse, so the map must be a bijection. Additionally, the map has to allow any complex analysis we do in the open subset of $\mathbb{C}$ to transfer back to the space in a way that "preserves the topological structure." In your algebra course you saw the idea of an *isomorphism*, a bijective map from one group (or ring or field) to another group (or ring or field) which preserves the structures of the group (or ring or ... you get the idea!). For groups, to "preserve the structure" means we get the exact same result whether we multiply elements of the preimage and then apply the map to the product or apply the map to elements of the preimage and then multiply the results in the image. In symbols, if f is the map, and $a, b \in G$, then $f(ab) = f(a)f(b)$. This (together with the bijective condition) is enough to ensure that isomorphic groups have the same cardinality and that they share other important characteristics like orders of their elements or abelian-ness. As another example of preserving structure, in linear algebra, you learned that linear transformations corresponding to orthogonal matrices preserve distance.

In topology, the structure-preserving bijective maps are called *homeomorphisms*. To "preserve structure" here means preserving the topological structure, so the map must send open sets (and only open sets) to open sets. Look for a formal definition in Sect. 2.8. Thus, our definition for the maps which give the local approximations of the surface by open subsets of the complex plane is:

Definition 2.4 Given a topological space X, a point p on X, a neighborhood U containing p, and an open subset $W \subseteq \mathbb{C}$, then a homeomorphism $\phi_U : U \to W \subseteq \mathbb{C}$ is called a **complex chart**.

One other technicality is if two neighborhoods U and V around different points intersect, the two complex charts ϕ_U and ϕ_V should behave similarly on that intersection. We would get into trouble in later arguments if we had to worry about which chart we were picking and if those choices affected the outcome. Remember, our goal is to be able to do analysis on the open subsets of $\mathbb{C}$ and then lift that back up to the surface X via the inverse of the complex chart. To make two charts compatible with one another, doing the lift back to X and then applying a different chart to get to a (possibly different) open subset of $\mathbb{C}$ should preserve analysis properties. We specify that the map $\phi_V \circ \phi_U^{-1}$ should be a *holomorphic map*, one that is complex differentiable at each point and, as such, preserves the analysis properties. This leads to the last two definitions we need before we can properly define a Riemann surface.

Definition 2.5 Two complex charts ϕ_U and ϕ_V are called **pairwise compatible** if $\phi_V \circ \phi_U^{-1}$ is holomorphic. A set of pairwise compatible complex charts so that each point in X is in at least one of the domains of the charts is called a **complex atlas** for X.

Riemann surfaces are topological spaces equipped with a complex atlas, and with a couple of other small restrictions which ensure the spaces behave "reasonably."

Definition 2.6 A topological space X which is Hausdorff, connected, and second countable, and which has a complex atlas defined on it is called a **Riemann surface**.

You can find the definitions for *Hausdorff*, *connected*, and *second countable* in the definition section. As you are first grappling with the concept of a Riemann surface, do not worry too much about these additional definitions. Most "nice" topological spaces, and all the examples in this paper, satisfy them.

Example 2.6 Consider $\mathbb{R}^2$, and fix any element $(x, y) \in \mathbb{R}^2$. Then the map $\phi : \mathbb{R}^2 \to \mathbb{C}$ sending $(x, y) \to x + iy$ is a homeomorphism, and a complex chart, and so $\mathbb{R}^2$ is a Riemann surface with the atlas $\{\phi\}$. □

Example 2.7 Any open subset of the complex plane is a Riemann surface, by defining one complex chart from the open subset projecting onto itself. For example, the *open unit disk* $D = \{z \in \mathbb{C} \mid |z| < 1\}$ is a Riemann surface with $\phi : D \to D \subseteq \mathbb{C}$ defined as $\phi(z) = z$, and atlas $\{\phi\}$. A similar argument follows for the *upper half plane* $\mathbb{H} = \{z \in \mathbb{C} \mid \mathrm{Im}(z) > 0\}$. □

Example 2.8 Sometimes a non-example can be just as enlightening. Consider the chart from Example 2.6, but restrict it to the open set of elements $(x, y) \in \mathbb{R}^2$ so that $x > -1$. Call that chart ϕ_{U_1}, and consider a second chart ϕ_{U_2} defined as

$\phi_{U_2}((x, y)) = x - iy$ on the open set of elements $(x, y) \in \mathbb{R}^2$ so that $x < 1$. In order for $\{\phi_{U_1}, \phi_{U_2}\}$ to be an atlas on $\mathbb{R}^2$, the two charts must be compatible on the intersection of the open sets, i.e., the open strip between $x = -1$ and $x = 1$. But on this intersection, $\phi_{U_2} \circ \phi_{U_1}^{-1}$ sends $x + iy$ to $x - iy$, which is not a holomorphic map. So $\{\phi_{U_1}, \phi_{U_2}\}$ is not an atlas on $\mathbb{R}^2$.

How do we know that the map sending $x + iy$ to $x - iy$ is not holomorphic? Complex functions can be described as having a real part and an imaginary part which are themselves real functions. There is a set of differential equations relating these real functions, called *the Cauchy-Riemann equations*, that are satisfied precisely when the corresponding complex function is holomorphic. In effect, the Cauchy-Riemann equations give a real analysis condition for when a complex function is differentiable. For our example, $\phi_{U_2} \circ \phi_{U_1}^{-1}$ can be written as $u(x, y) + v(x, y)i$ where $u(x, y)$ is a real function in two variables sending (x, y) to x, while $v(x, y)$ sends (x, y) to $-y$. One of the Cauchy-Riemann equations is $\frac{\partial u}{\partial x} = \frac{\partial v}{\partial y}$, but in our case, this equation is not satisfied as $\frac{\partial u}{\partial x} = 1$ but $\frac{\partial v}{\partial y} = -1$. And so $\phi_{U_2} \circ \phi_{U_1}^{-1}$ is not holomorphic. □

Before giving another example, we pause here to discuss what it means for two Riemann surfaces to be equivalent.

Definition 2.7 Given Riemann surfaces X and Y, a map $f : X \to Y$ is **holomorphic** at a point p if there is a chart ϕ_U of X with $p \in U$ and a chart ϕ_V of Y with $f(p)$ in V, so that $\phi_U \circ f \circ \phi_V^{-1}$ is holomorphic. If such charts exist for all $p \in X$, then f is a **holomorphic map**.

This definition is a good example of the motivation for defining charts as we do. There is no way to define a derivative on an arbitrary surface. But, because Riemann surfaces look like complex planes locally, we can use the definition of holomorphic in $\mathbb{C}$ and translate it back to X using charts, giving us a definition of a holomorphic map on X. Notice also that pairwise compatibility of charts is important in these definitions so that we do not need to worry about the choice of ϕ_U and ϕ_V. If we choose different charts for points in X, the holomorphicity still carries through.

Example 2.9 We can define a holomorphic map between the two Riemann surfaces in Example 2.7. You should check that the map $f : \mathbb{H} \to D$ with $f(z) = \frac{z-i}{z+i}$ actually sends elements in the upper half plane to elements in the open disk and that f is a holomorphic map. There is also a holomorphic inverse map $f^{-1} : D \to \mathbb{H}$ defined as $f(z) = -i\frac{z+1}{z-1}$ and so $\mathbb{H}$ and D are *isomorphic* Riemann surfaces. □

Another popular example of a Riemann surface, and one we will use later in the paper, is the *Riemann sphere*.

Example 2.10 Take the unit sphere in $\mathbb{R}^3$, defined as $S^2 = \{(x, y, z) \in \mathbb{R}^3 \mid x^2 + y^2 + z^2 = 1\}$, call the points $(0, 0, \pm 1)$ the *north and south poles*, and consider the (x, y) plane which, as we just saw in Example 2.6, is homeomorphic to $\mathbb{C}$. To ease our conversation, consider the points (x, y) written as $x+iy \in \mathbb{C}$. Through a process

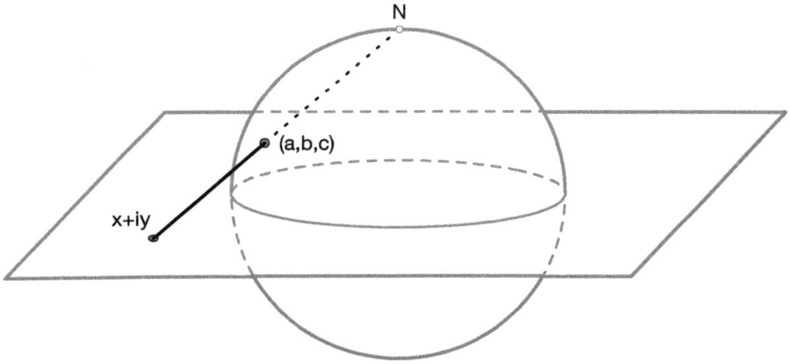

Fig. 2.3 The line forming the bijection between points on S^2 minus the north pole and points on the complex plane

called *stereographic projection*, we will define two charts from $S^2 - \{(0,0,1)\}$ and $S^2 - \{(0,0,-1)\}$ to the complex plane (represented as the (x, y) plane).

Begin with the north pole (labeled as N in Fig. 2.3). Any line through the north pole will hit the sphere in exactly one place and will hit the complex plane also in exactly one place. These lines, therefore, define a bijection from points on $S^2 - \{(0,0,1)\}$ to points on $\mathbb{C}$. In your multivariable calculus class, you learned how to define lines in $\mathbb{R}^3$, so the line going through $(0, 0, 1)$ and intersecting the sphere at a point $(a, b, c) \in S^2$ can be parametrized as $\ell(t) = (0, 0, 1) + (a, b, c - 1)t$. Where will this line intersect the (x, y) plane? When the z coordinate is 0, or, solving for t in ℓ, when $t = 1/(1 - c)$. Simply plug this value of t into $\ell(t)$ to get a map $\phi_N : S^2 \to \mathbb{C}$ so that $\phi_N(a, b, c) = \frac{a}{1-c} + \frac{b}{1-c}i$.

The map ϕ_N is a bijection. Its inverse map ϕ_N^{-1} is created in much the same way as we created ϕ_N. Take some $x + iy \in \mathbb{C}$ and form the line from the north pole to $x + iy$ (or really from its equivalent point in $\mathbb{R}^3$: $(x, y, 0)$), and determine where that line intersects the sphere S^2. You will get a formula

$$\phi_N^{-1}(x, y) = \left(\frac{2x}{x^2 + y^2 + 1}, \frac{2y}{x^2 + y^2 + 1}, \frac{x^2 + y^2 - 1}{x^2 + y^2 + 1} \right).$$

To make this map a chart, we do need to confirm that it is homeomorphic. How do we know ϕ_N and its inverse are homeomorphic (i.e., continuous)? We can say that both maps are compositions of known continuous functions (polynomials and rational functions) and thus are continuous.

Now, go through an almost identical argument yourself using the south pole to create a second complex chart $\phi_S : S^2 \to \mathbb{C}$ so that $\phi_S(a, b, c) = \frac{a}{1+c} - i\frac{b}{1+c}$. Before we can declare the set $\{\phi_N, \phi_S\}$ to be an atlas on S^2, we must ensure the two charts are pairwise compatible. The composition $\phi_S \circ \phi_N^{-1} : \mathbb{C} \to \mathbb{C}$ sends z to $1/z$, which is a well-known holomorphic map. Once more I leave the thrilling calculations to you, dear reader. □

The Riemann sphere is often called $\mathbb{C} \cup \{\infty\}$, since S^2 minus the north pole is homeomorphic to $\mathbb{C}$, and we can think of N as the limit point of any unbounded sequence in $\mathbb{C}$, hence a "point at infinity." We will discuss the process of adding a point in the next section, as it connects to a particularly nice form that all of the Riemann surfaces in my research have. Then in Sect. 2.4.2, we return to examples of Riemann surfaces by considering a special family I use in my research.

2.4.1 Compactness

Sets which are closed and bounded in $\mathbb{R}$ have nice properties. As just one example, the Extreme Value theorem says that on a closed and bounded subset of $\mathbb{R}$, any continuous function attains an absolute maximum and minimum. We would like to extend the ideas of closed and bounded sets beyond $\mathbb{R}$. This is where a property called *compactness* comes in. In your introductory analysis course, you may have seen a definition for compactness involving open covers. The topological spaces we care about in this paper are special cases of objects called *metric spaces*, and in metric spaces, this definition of compactness is equivalent to the following one, which will be easier to work with in our setting.

Definition 2.8 Topological spaces in this paper are **compact** if every sequence has a convergent subsequence.

Is the real line compact? (If compactness is a generalization of closed and bounded sets then the answer had better be no!) The sequence $(1, -1, 1, -1, 1, -1, \ldots)$ in $\mathbb{R}$ does not converge but does have a convergent subsequence consisting of every even or every odd entry. The sequence $\left(1, 1, 2, \frac{1}{2}, 3, \frac{1}{3}, 4, \frac{1}{4}, \ldots\right)$ in $\mathbb{R}$ has a subsequence that converges to 0. But not all sequences have convergent subsequences. Take the sequence of integers $(1, 2, 3, 4, \ldots)$. This sequence most definitely does not have a convergent subsequence as the limit of every subsequence will still be ∞. So $\mathbb{R}$ is not a compact Riemann surface.

The failure of compactness in $\mathbb{R}$ boils down to a problem with unbounded sequences as, by a result you probably saw in your real analysis course called the Bolzano-Weierstrass theorem, any bounded sequence $\mathbb{R}$ will have a convergent subsequence. Unbounded sequences in $\mathbb{R}$ all have a subsequence that heads to $\pm\infty$ and that subsequence is the crux of the problem. If we could just add a point called infinity to $\mathbb{R}$, and define that point to be the limit of the unbounded sequences, then we would have a compact space. Geometrically, imagine identifying $\pm\infty$ as one new point and, by adding that one point, producing a circle. We have thus created a compact space $S^1 = \{(x, y) \in \mathbb{R}^2 \mid x^2 + y^2 = 1\}$ simply by adding one point to $\mathbb{R}$. The process of turning a surface which is not compact into a compact one by adding points has the fun name of "compactification"!

The complex plane is also not compact for similar reasons: The sequence of integers $(1, 2, 3, 4, \ldots)$ does not have any convergent subsequences. We can fix this (create a compactification of $\mathbb{C}$) by adding a point at infinity. In Example 2.10, we described just such an object, called the *Riemann sphere*.

Since intuitively the idea of compactness should be connoted with the ideas of closed and bounded, compact spaces should not go off to infinity (like the upper half plane $\mathbb{H}$ from Example 2.7) but instead should fold in on themselves, and they should not be open (like the unit disk D also in Example 2.7). When you think of compact Riemann surfaces, you can have in your mind hollow shapes in $\mathbb{R}^3$ like a sphere or a torus (holed donut) or a many holed torus.

2.4.2 Curves

The family of compact Riemann surfaces I work with are *curves*. You are well acquainted with curves, at least over $\mathbb{R}$. For example, the unit circle, defined as the set of solutions in $\mathbb{R}^2$ to the equation $x^2 + y^2 = 1$, is a curve. Notice we could also define it as solutions to $f(x, y) = 0$ where $f(x, y) = x^2 + y^2 - 1$. Similarly for curves over $\mathbb{C}$, take some polynomial $f(z, w) \in \mathbb{C}[z, w]$, the polynomial ring in two variables with coefficients in $\mathbb{C}$, and assume the degree of f is greater than 0 to avoid some pathological cases. Let

$$X = \{(z, w) \in \mathbb{C}^2 \mid f(z, w) = 0\}. \tag{2.1}$$

This set X is a (plane) **curve**.

Near the beginning of Sect. 2.4, we describe several analytic advantages to studying the complex plane. In addition, the complex numbers are *algebraically closed*, which means any polynomial with coefficients in $\mathbb{C}$ has all its roots in $\mathbb{C}$. While the polynomial $x^2 + 1$ has no roots in $\mathbb{R}$, if we consider the same polynomial over $\mathbb{C}$, we see its two roots: $\pm i$. Therefore, we study curves that are defined by polynomials with coefficients in $\mathbb{C}$ both to align analytically with Riemann surfaces, and also so we do not need to worry about whether all roots of the polynomials exist in the field we are working in.

To define a Riemann surface structure on the curve, we must add some extra conditions. First, we want the curve to be connected, one of the conditions in our definition of Riemann surfaces. One way to guarantee connectedness is to insist that the polynomial $f(z, w)$ above be irreducible. The other condition is that we want the curve to be *smooth*, meaning it has a defined tangent at all points. From the definition of the tangent to a curve, this means the partial derivatives of f exist for each point on the curve and are never simultaneously zero. For the rest of the paper, if I use the word "curve," assume I mean a smooth, connected curve.

How do we define complex charts for curves? For each point $p \in X$, we want an open neighborhood U containing p to exist so that the projection map $\pi_U : U \to W \subseteq \mathbb{C}$ sending (z, w) to z is a homeomorphism, and thus a complex chart. Projection is continuous, since if we take an open set around a point z_0, its preimage under the projection will also be an open set.

When will π_U be injective? The answer to this question requires the Implicit Function theorem. This theorem is difficult to prove, but you have certainly used versions of it. Take our original example of a curve, the unit circle defined as $f(x, y) = x^2 + y^2 - 1 = 0$ which is not a function in $\mathbb{R}$. However, we can define subintervals on which $f(x, y) = 0$ can be described as a function. For instance, the top half of the circle is equivalent to $g(x) = \sqrt{1 - x^2}$ which is a function. Or for the bottom half, we can use $g(x) = -\sqrt{1 - x^2}$. The only points in $\mathbb{R}^2$ for which we cannot use either $g(x)$ defined above are the points $(\pm 1, 0)$. No matter what neighborhood you take around either of these two points, one x value will always correspond to two y values, and so there is no way to define a function $g(x)$ which represents the circle in a neighborhood around $(\pm 1, 0)$. However, the same is not true if we switch the roles of x and y. Near the points $(\pm 1, 0)$, each y value corresponds to exactly one x value, and we can instead use the functions $g(y) = \pm\sqrt{1 - y^2}$ to locally define the circle. This is the idea of the Implicit Function theorem. For curves in $\mathbb{C}^2$ given by an equation $f(z, w) = 0$ and with special conditions on the partial derivatives, we are able to define holomorphic functions that are equal to the defining equation of the curve in small neighborhoods, although these functions may not all be defined with the same variable.

Back to the quest for complex charts in X. Take a curve X defined as in (2.1) and any point $p = (z_0, w_0)$ on it. Since the curve is smooth, we know that at least one partial derivative is nonzero at the point p. Without loss of generality, say it is the derivative with respect to w_0. Since that partial derivative is nonzero, we can apply the Implicit Function theorem, to conclude there is a neighborhood $U \subseteq X$ containing the point p, so that every point in U is of the form $(z, g(z))$ for some function $g(z) \in \mathbb{C}[z]$. The key here is each z corresponds to just one w value in U which ensures that the map π_U has an inverse sending z to $(z, g(z))$, and so π_U is a bijection on U. We already said π_U is continuous, and the Implicit Function theorem tells us that its inverse is continuous; hence π_U is a homeomorphism. If the partial with respect to w_0 had been 0, then we would define the projection $\pi_U : U \to W \subseteq \mathbb{C}$ sending (z, w) to w and with inverse sending w to $(g(w), w)$ for a function $g(w)$ which exists by the Implicit Function theorem.

Now we have charts, the π_U, but are they pairwise compatible? If π_U and π_V are two charts so that $U \cap V \neq \emptyset$ and both are defined as projection on the same variable, then $\pi_V \circ \pi_U^{-1}$ is the identity map which is certainly a holomorphic map. If, say, π_U is projection onto the first variable, and π_V is projection onto the second variable, then $\pi_V \circ \pi_U^{-1}(z) = \pi_V((z, g(z))) = g(z)$ i.e. $\pi_V \circ \pi_U^{-1}$ is equal to the function $g(z)$ which is holomorphic by the Implicit Function theorem. Thus the curve X is a Riemann surface with complex charts ϕ_U.

I have hidden until now one flaw with our discussion about curves as Riemann surfaces. The curves we define above are called *affine curves*, and the corresponding Riemann surface is not compact. In order to make the Riemann surface compact, we instead have to work with objects called *projective curves*. These are curves defined over *projective space* which, similar to our conversation about compactness in Sect. 2.4.1, is a space formed by adding a point at infinity. This is a tale for another time, but if you would like to read more about projective space, I recommend [4, Chapter 8].

It is also true that every compact Riemann surface is a projective curve, but that is not easy to prove! An overview of this result, including several references on where to find more details, is in Theorem 4.16.1 in [13]. The idea that certain analytic objects (compact Riemann surfaces in this case) are equivalent to certain algebraic objects (projective curves in this case) is a very deep connection between algebraic and analytic geometry, with the catchy name of GAGA, the acronym for the French title of the main paper connecting these ideas: Serre's "Géometrie Algébrique et Géométrie Analytique" [24]. The beauty of this connecting theory is that now we can bring both algebraic and analytic tools to bear on problems about these objects, and this opens up many more areas for exploration and understanding. I doubt GAGA has anything to do with the musician Lady Gaga, although I like to imagine a world where she chose her name in part because of this area of mathematics!

2.5 Groups Acting on Riemann Surfaces

Suppose we have a group G acting on a Riemann surface X. Recall from Sect. 2.3 that the orbits of a group action partition the set. For a given group G acting on X, this means the points on X may be collected into compartments, based on what orbit they are in. We call the set of orbits (the compartments) X/G. Geometrically, what is this set of orbits? Instead of a group action, first think more generally of equivalence relations on the points on X. As you may recall from Sect. 2.3, a group action produces an equivalence relation.

As an example, take the closed disk $D = \{z \in \mathbb{C} \mid |z| \leq 1\}$ and define an equivalence relation as z_1 and z_2 in D are equivalent if they are both on the boundary, or if $z_1 = z_2$. The equivalence relation partitions D into the boundary (as one compartment) and each point on the interior as its own compartment. What happens geometrically if we identify all the points on the boundary, i.e., what does the set of partitions look like? Visually, imagine shifting the disk into three dimensions by pulling up on its boundary, creating a bowl shape. Then push together the rim of the bowl to one point, forming a sphere. This tells us the set of partitions for this equivalence relation is (homeomorphic to) the sphere S^2.

The same idea transfers over to the special case of a group acting on a Riemann surface. Geometrically, we glue together into one point all the points in X which are in the same orbit (not so unlike the idea of *tessering* [15]), and do that for each orbit, which suggests we can put a topological structure on X/G too. The set of orbits X/G has the *quotient topology* inherited from the topology of X via the natural map $\pi : X \to X/G$ sending any point p to its orbit O_p. The precise definition of quotient topology is found in the definition section. On a first reading, it is fine to take on faith that X/G is a topological space.

Here we are interested in the fact that X/G is also a Riemann surface. To make things a bit nicer, we assume the action of G on X is *faithful*, i.e., the intersection of all the stabilizers is trivial (see Sect. 2.3). This condition is not very restrictive, as it turns out that a non-faithful action of G on X can be represented by a faithful action (for the experts: quotient G out by the kernel of the action). We also assume, to preserve analytic structure, that each of the $\varphi_g : X \to X$ defining the group action is a *holomorphic map on* X (we defined a holomorphic map between Riemann surfaces in Sect. 2.4).

What are charts for X/G? For each point in X/G (an orbit O_p where $p \in X$), we need to find an appropriate neighborhood $U \subseteq X/G$ around O_p and a homeomorphic map ψ_U from U to an open subset of $\mathbb{C}$, so that the different ψ_U are pairwise compatible. Where could we get such a map ψ_U? Your instincts may be screaming at you to consider the charts we already have from the atlas of X, call them ϕ_V where V is a neighborhood of a point in X. We also have at our disposal the projection $\pi : X \to X/G$. Maybe we can use some sort of inverse of π composed with one of the charts of X to get a map $\phi_V \circ \pi^{-1} : U \to W \subseteq \mathbb{C}$?

The issue we encounter is that to properly define an inverse of π, we need a one-to-one map, at least when restricted to some open subset of X. For an open subset $V \subseteq X$, call the restricted map $\pi|_V : V \to X/G$. If $\pi|_V$ were one-to-one, this would mean whenever $\pi|_V(p_1) = \pi|_V(p_2)$ then $p_1 = p_2$. And what does it mean if $\pi|_V(p_1) = \pi|_V(p_2)$? It means p_1 and p_2 are in the same orbit under the action of G, or there is some $g \in G$ so that $\varphi_g(p_1) = p_2$. Thus, $\pi|_V$ will be one-to-one if V contains no distinct elements from the same orbit.

To execute our plan of using charts of X together with π to define charts on X/G, we therefore need to find, for each point in X, a neighborhood V with not more than one element from each orbit in it. It turns out that when a group G acts on a Riemann surface X faithfully and holomorphically, we can always find such V for any p which has trivial stabilizer. Construction of these neighborhoods is a bit technical, but the interested reader can find all the details in [17, Proposition III.3.3]. The Hausdorff property of Riemann surfaces plays a role here, as we can separate each of the distinct points in the orbit into different neighborhoods.

Equipped with this one-to-one map $\pi|_V$, we create the charts in X/G. Start with some point in X/G, call it O_p. Form the neighborhood V in X of p as discussed above, so that the quotient map π restricted to V is a homeomorphism onto a neighborhood U of O_p in X/G. Now, because X is a Riemann surface, there is a chart of X defined as $\phi_V : V \to W \subseteq \mathbb{C}$ (we may need to shrink V to apply the chart). And so $\psi_U = \phi_V \circ \pi|_V^{-1} : U \to W$ is a chart for the neighborhood U of O_p. It is homeomorphic, since it is a composition of homeomorphic functions. Why are these charts pairwise compatible? Write out $\psi_{U_2} \circ \psi_{U_1}^{-1}$ as a composition of ϕ_{V_i} and $\pi|_{V_i}$, and observe that the projection maps cancel out, leaving only $\phi_{V_2} \circ \phi_{V_1}$ which is holomorphic, since the ϕ_{V_i} are pairwise compatible.

The whole argument above does not quite work for points p with nontrivial stabilizers, but the general idea is still the same. Again, see [17, Proposition III.3.3] for the case of nontrivial stabilizers. Thus, if a group G acts faithfully and holomorphically on a Riemann surface X, the set of orbits X/G will also be a Riemann surface.

One final point on the way out of this subsection. The image via a continuous function of a compact space is again a compact space, and so X/G will be compact whenever X is because π is continuous.

2.5.1 *Automorphisms*

In the previous section, we talked abstractly about an arbitrary group acting on a Riemann surface. But what sort of groups act on Riemann surfaces? Geometrically, the action will be a shuffling of the points in X, so how can a group represent shuffling of points? Consider curves, which we saw are Riemann surfaces in Sect. 2.4.2. Suppose we have a curve X defined by the equation $f(x, y) = y^2 - x^6 - x^3 - 1 = 0$. If there is one solution (x_0, y_0) to the equation $y^2 = x^6 + x^3 + 1$, pause and think about what other solutions must then exist in $\mathbb{C}$.

Perhaps you came up with the solution $(x_0, -y_0)$. With a little more work, you may have found the solution $(\omega_3 x_0, -y_0)$ where ω_3 is a cube root of unity, that is $\omega_3 = \frac{-1}{2} + \frac{\sqrt{3}}{2}i$, or the solution $\left(\frac{1}{x_0}, \frac{y_0}{x_0^3}\right)$ as long as $x_0 \neq 0$. These additional solutions lead to two new maps from X to X. Call the map r which sends any point (x_0, y_0) to $(\omega_3 x_0, -y_0)$, and call the map s which sends each point (x_0, y_0) to $\left(\frac{1}{x_0}, \frac{y_0}{x_0^3}\right)$, except for the point $(0, 1)$ which is fixed by s. These two maps define ways to permute the points on this curve. If we want a group action, we simply need to create a group generated by these two elements, under the operation of composition. For instance, $r \circ r = r^2 : X \to X$ sends (x_0, y_0) to $(\omega_3^2 x_0, y_0)$, and $r^3 : X \to X$ sends (x_0, y_0) to $(x_0, -y_0)$ (hey, that looks familiar!). Figure out higher powers of r and what the inverse of r is. Now do the same thing with s. Play around to see what $r \circ s$ and $s \circ r^{-1}$ are. Can you identify the group generated by r and s? (Hint: we probably asked you about $r \circ s$ and $s \circ r^{-1}$ for a reason!) This group is (isomorphic to) the dihedral group on 6 objects, or D_6. It turns out that this is the full automorphism group of this curve, but that is not easy to see.

Now we have an explicit example of how the dihedral group acts on the curve (Riemann surface) defined by the equation above. This example is a specific case of a much more general construction, as I am sure you can imagine finding similar maps given a different equation. Remember that when we studied X/G in the previous section, we restricted to actions defined by holomorphic maps. Well, the maps above are holomorphic maps with holomorphic inverses.

Definition 2.9 An **automorphism** of a Riemann surface X is a holomorphic map from X to itself which is also a bijection, and with an inverse that is also holomorphic. The set of automorphisms form a group under function composition called the **automorphism group**.

The automorphism group of a Riemann surface is one example of a group which can act on a Riemann surface. These groups are particularly relevant to my research, and I talk about them in Sect. 2.6. While it is not obvious, the automorphism groups of compact Riemann surfaces are all finite and in fact bounded in size by an invariant of the surface called the *genus* of X. Topologically, the genus measures how many holes a surface has, and for curves, it roughly measures how big the degree of the defining polynomial is. The formal result is *Hurwitz's Theorem* which says that a Riemann surface X of genus g has an automorphism group G so that $|G| \leq 84(g-1)$ [17, III.3.9]

2.5.2 Riemann Existence Theorem

So far, we have determined that if G is a group acting on a compact Riemann surface, then X/G is also a compact Riemann surface. We saw that automorphism groups of curves give examples of groups acting on Riemann surfaces, and for compact Riemann surfaces (remember, these are the same as projective curves), the groups are finite. Can we be more explicit about which finite groups are automorphism groups of some compact Riemann surface? Indeed we can, and the crux of this theory is the beautiful Riemann Existence theorem [9], a version of which we state here.

Each action of some group G on a compact Riemann surface X has a list of integers $[m_1, \ldots, m_r]$ associated with it which satisfy a formula called the Riemann-Hurwitz formula [2, Lemma 3.13]. This formula relates the size of G, the m_i, and the genus of X and X/G. Recall from the introduction that for any $g \in G$, the notation $o(g)$ represents the order of g.

Theorem 2.1 *Given a list of integers $[m_1, \ldots, m_r]$ satisfying the Riemann-Hurwitz formula, there is a finite group G and elements $g_1, \ldots, g_r \in G$ so that*

1. *for all i, $o(g_i) = m_i$,*
2. *the g_i together generate G, and*
3. *the product $g_1 g_2 \cdots g_r$ is the identity of G, if and only if G is the automorphism group of some compact Riemann surface X. In this case, the quotient X/G is isomorphic to the Riemann sphere.*

Example 2.11 For the alternating group A_6, the elements $g_1 = (1\ 3)(4\ 6)$, $g_2 = (1\ 5\ 6\ 4)(2\ 3)$, and $g_3 = (1\ 2\ 3\ 4\ 5)$ satisfy the Riemann Existence theorem for $m_1 = 2$, $m_2 = 4$, and $m_3 = 5$. The elements $g_1 = (1\ 2\ 3)(4\ 5\ 6)$, $g_2 = (1\ 4\ 2\ 3)(5\ 6)$, and $g_3 = (1\ 2\ 3\ 4\ 5)$ satisfy the Riemann Existence theorem for $m_1 = 3$, $m_2 = 4$, and $m_3 = 5$. □

A topological proof of this theorem exists which is constructive (see [27, Theorem 4.27 and Theorem 4.32], although be warned that this is a very advanced book).

What is so powerful about the Riemann Existence theorem is that it connects topology, group theory, and complex geometry in the study of compact Riemann surfaces, giving us many powerful tools in each of these disciplines to attack problems with. In my humble opinion, this is where the most beautiful mathematics exists, along boundaries of different disciplines within mathematics. For my research, this theorem gives me a way to find explicit automorphism groups of compact Riemann surfaces by using group theory.

I am sure you are wondering why the conditions on the elements g_i in the theorem are what they are. The map $\pi : X \to X/G$ is more than just a quotient map; it is also a *covering*, branched at r places (r being the number of g_i in Theorem 2.1). If you have seen enough covering space theory, the g_i are generators of the monodromy group. As permutations, they describe what happens to the preimages of the fixed base point in X/G when loops starting and ending at that base point and traveling once around the ith branch point are lifted to X. This idea will not be explored further here, but if you are interested in learning more, a good graduate level reference for covering space theory and group actions related to them is Chapter 11 and 12 in [14]. For the specific situation discussed here, we refer you to [9] or [2] or [27].

You may also be wondering why the Riemann Existence theorem specifies that X/G is the Riemann sphere and if we could generalize to other Riemann surfaces. The answer is definitely yes, but things get more complicated if we consider other surfaces. For the experts in the room, if X/G has genus g_0, we have to add $2g_0$ generators of G and conditions on the products of commutators of these generators, corresponding to the fundamental group of a genus g_0 surface.

2.6 Contributions

My research path has led me to study objects called *Jacobian varieties*, and how they factor into smaller pieces. Much like the prime factorization trees we all remember from grade school (see Fig. 2.4), many mathematical objects can be broken down into a product of simpler pieces. These pieces are often easier to study, and they give us additional information about the original object. In your linear algebra course, you probably talked about decomposing matrices. For example, the QR decomposition is a way to write a real square matrix as the product of an orthogonal matrix and an upper triangular matrix. This decomposition lets us better understand how the linear transformation acts on vectors. You also saw an example in your introductory algebra class, in the form of *direct products*. If C_n represents the cyclic group of order n, then $C_6 \cong C_3 \times C_2$ and, perhaps more surprisingly, $D_6 \cong D_3 \times C_2$.

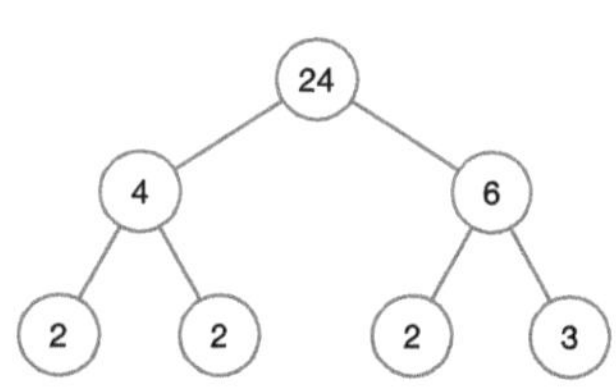

Fig. 2.4 The factorization of the integer 24 into a product of prime integers

While defining Jacobian varieties carefully is sadly beyond the scope of this paper (take a graduate level course in algebraic geometry and then read [16] or the relevant parts of [10]), I can give you a bit of information about them. First what are *varieties*? They are solutions to a system of polynomial equations. Curves (see Sect. 2.4.2) are a special example of varieties defined by one equation in two variables. Second, Jacobian varieties are *abelian varieties* which means they have a group structure on their points (not the group action we talked about, just an honest to goodness group formed naturally from the points). Finally, each Jacobian variety is connected in a canonical way to a curve, as the group structure of the Jacobian variety gives us information about the points on the corresponding curve.

Jacobian varieties can be viewed as a generalization of *elliptic curves*. If you have already been introduced to elliptic curves, you know that they have a beautiful and natural group structure on their points. If you have not seen elliptic curves before, or would like a refresher, I highly recommend [25]. The group structure of elliptic curves is very powerful, and the Jacobian variety group structure allows us to recover some of this power for curves which do not have a natural group structure on their points.

Fix your favorite (projective) curve X. Remember that projective curves are compact Riemann surfaces, so our discussions of group actions apply to the curve X. The symbol JX represents the Jacobian variety of this curve. My interest in factoring Jacobian varieties started with my Ph.D. thesis, where I explored how to leverage the group action of automorphisms of X to decompose JX into products of smaller abelian varieties [20] (here smaller means smaller topological dimension). I looked for examples where the Jacobian variety factored into the product of one elliptic curve, or $JX \sim E^g$, where E^g means the product of g copies of some elliptic curve E. Elliptic curves play a role similar to prime numbers in integer factorization: they are the "smallest" abelian varieties (they are all dimension 1) and so cannot be factored more. The $\sim$ symbol represents an *isogeny*, a map which is not quite an isomorphism, but close enough for most purposes, i.e., it preserves the structure we care about.

The decomposition technique requires a subject called representation theory (groups acting on vector spaces) as well as some more advanced topology. If you would like to read more about representation theory assuming only an undergraduate background, I recommend [12]. Side note: As an undergraduate, I was not overly fond of linear algebra, so I happily avoided taking a representation theory course in graduate school, convinced I would never, ever, ever need that material. The fact that I frequently use this topic in my research now is more evidence that the universe has a demented sense of humor, and that you never know where your mathematical life will take you!

In Sect. 2.5.1, we mentioned bounds for the size of the automorphism group of any curve of a fixed genus. There are families of groups which attain this bound. For example, the *projective special linear groups* $\mathrm{PSL}(2, q)$ are a family of groups defined by taking 2×2 matrices with coefficients in the finite field with q elements and with determinant 1 and forming the quotient group by the center, the subgroup $\{\pm I\}$. These groups are the automorphism groups of a family of curves and of maximal possible order for the genera of the curves in that family. Since the representation theory of these groups is well known, finding the Jacobian variety decomposition of the corresponding curves is a tractable problem. Two undergraduates worked with me through a summer program at Grinnell College and determined this decomposition. [8]. Here is the key result they proved.

Theorem 2.2 (Theorem 10 [8]) *Fix a curve X with automorphism group $PSL(2, q)$, so that q is odd and $q > 27$, and which is of maximal possible order for the genus of X. Let A_x^y be the product of y copies of an abelian variety of dimension x, while u, v, and w are values depending on q* mod 168, *and let $\phi(x)$ be the Euler phi function.*

When $q \equiv 1 \bmod 4$

$$JX \sim A^{q}_{\frac{q-u}{84}} \times A^{\frac{q+1}{2}}_{\frac{q-v}{84}} \times \prod_{\substack{d \mid \frac{q-1}{2} \\ d < \frac{q-5}{4}}} A^{q+1}_{\frac{\phi\left(\frac{q-1}{2d}\right)\cdot(q-w)}{168}} \times \prod_{\substack{d \mid \frac{q+1}{2} \\ d < \frac{q-1}{4}}} A^{q-1}_{\frac{\phi\left(\frac{q+1}{2d}\right)\cdot(q-w)}{168}}.$$

When $q \equiv -1 \bmod 4$

$$JX \sim A^{q}_{\frac{q-u}{84}} \times A^{\frac{q-1}{2}}_{\frac{q-v}{84}} \times \prod_{\substack{d \mid \frac{q-1}{2} \\ d < \frac{q-3}{4}}} A^{q+1}_{\frac{\phi\left(\frac{q-1}{2d}\right)\cdot(q-w)}{168}} \times \prod_{\substack{d \mid \frac{q+1}{2} \\ d < \frac{q-3}{4}}} A^{q-1}_{\frac{\phi\left(\frac{q+1}{2d}\right)\cdot(q-w)}{168}}.$$

In particular, these Jacobian varieties always factor into smaller pieces (they never behave like prime numbers). I reproduce this result to give you an idea of the kind of theorems we prove, not because I expect you to have a deep understanding of these Jacobian varieties simply by staring at their decompositions. If $q \leq 27$ and PSL$(2, q)$ is the automorphism group of a Riemann surface, then the only options for q are values in the set $\{7, 8, 13, 27\}$. In the first three cases, the Jacobian variety decomposes completely into a product of elliptic curves. If $q = 27$, we can still decompose the Jacobian variety. It just has a slightly different form which comes from the fact that the conjugacy class structures of PSL$(2, 27)$ are not the same as for larger q (for the experts: there are two conjugacy classes of elements of order 3, whereas for larger q, there is only one).

Anita Rojas from Universidad de Chile and I have worked toward answering an open question from [6]. They ask if there are arbitrarily large curves, so that the Jacobian variety of the curve is *completely decomposable*, meaning it factors into all elliptic curves (not necessarily all the same curve). By "arbitrarily large," we mean of arbitrarily large genus. They produce examples of such curves up to genus 1297, but there are many gaps in their list of examples ("gaps" here means missing *genera*, my favorite plural word in mathematics).

To find more examples, we created a new technique to decompose more Jacobian varieties [22]. If G is the automorphism group of some curve X, the quotient X/G (which, remember, is also a Riemann surface) may not have any automorphisms which prevents us from being able to decompose the Jacobian of X/G using old techniques. Our new method uses the decomposition of JX to induce a decomposition of the Jacobian of X/G, the thitherto indecomposable Jacobian variety. We applied this technique to many examples, using the Riemann Existence theorem to determine automorphism groups, and found many new genera with completely decomposable Jacobian varieties.

Theorem 2.3 (Theorem 1 [22]) *For every integer g in the following list, there is a curve of genus g with completely decomposable Jacobian variety found using a group acting on a curve.*

*1–29, **30**, 31, **32**, 33, **34–36**, 37, **39**, 40, 41, **42**, 43, **44**, 45, **46**, 47, **48**, 49, 50, **51–52**, 53, **54**, 55, 57, **58**, 61, 62, **63**, 64, 65, **67**, **69**, **71**, 72–73, **79–81**, 82, **85**, **89**, **91**, **93**, **95**, 97, **103**, **105–107**, 109, **118**, 121, **125**, 129, **142**, 145, **154**, 161, 163, **193**, **199**, **211**, **213**, 217, **244**, 257, 325, 433*

The numbers in bold are new examples, the other examples are all different from those in [6].

In another paper, I explore the factorization of Jacobian varieties of a family of curves called *hyperelliptic curves* [21]. These curves have nice defining equations of the form $y^2 = f(x)$ with $f(x) \in \mathbb{C}[x]$, and their automorphism groups are well known which makes proving more general results about the Jacobian factorizations possible.

When I introduced the Riemann Existence theorem, you may have asked yourself some questions such as, "Which groups satisfy it? Which elements of those groups satisfy it?" An algorithm which can determine groups and elements satisfying the Riemann Existence theorem for a fixed genus exists [2], and complete lists have been determined through genus 48. This data is very useful not only for my research (as you can see above) but also for other research mathematicians as well. Enter the *L-functions and Modular Forms Database* (LMFDB), a huge collaborative project among over 80 mathematicians to centralize mathematical data, with a particular goal of making it accessible to any researcher, regardless of their prior computing experience. I have added the data from [2], as well as several other key pieces of data here: http://www.lmfdb.org/HigherGenus/C/Aut/. Adding additional data is an ongoing project, as there is a lot of interesting information about group actions on compact Riemann surfaces which may be included.

2.7 Future Work

To intentionally misquote a proverb, "A mathematician's work is never done." Anita Rojas and I have produced more evidence toward answering the question in [6] about whether there are arbitrarily large curves with completely decomposable Jacobians, but the question has not yet been answered. It is likely that in order to make progress on the general question we will need to move beyond group actions to find other ways to factor Jacobian varieties.

Similar to the family of groups PSL(2, q), the alternating groups A_n are also the automorphism groups of many curves with maximal possible order for their genus. The representation theory of these groups is more complicated than for PSL(2, q), but students working with me have made some small progress in understanding these decompositions.

Due to the nature of the decomposition technique for Jacobian varieties, we often do not know which abelian varieties the factors are. In fact, we rarely know equations defining even the original curves. Consider the statement of the Riemann Existence theorem. It says that if we find group elements with specific properties, then there is some curve which has that group as an automorphism group, but it does not give us a way to determine the equation of the corresponding curve. If we could find ways to determine which curves correspond to which automorphism group in higher genus, there are many interesting questions we could answer about the factors. Buzzwords for the experts: Do they have complex multiplication? What sort of torsion or rank do the elliptic curve factors have?

As I mentioned, there are a number of interesting questions related to Riemann's Existence theorem, and more generally to automorphism groups of curves. Right now, I am thinking about which lists of m_i produce examples for a given group. There are a number of combinatorial and group theoretic results that go into studying this question. A good reference to give you a sense of recent progress in this general area is a collection of research level papers [11]. If you are interested in open problems in the area, check out a survey paper with over 70 suggested problems which I wrote with my collaborators Allen Broughton and Aaron Wootton [3].

2.8 Definitions

This section contains formal definitions to tangential ideas discussed in this paper, on topics you may have not yet seen in your undergraduate education. The definitions are given roughly in the order they appear in the paper, starting in Sect. 2.4. Most are in the area of topology or complex analysis. A standard reference for undergraduate topology material is [18]. I learned from [1] and remember enjoying the book as a student. Just about any undergraduate complex analysis textbook will give you the basic information about the subject we use here. If you would like to get some nice intuition about complex analysis, I suggest checking out [19].

Definition 2.10 Given a set X, a **topology** on X is a collection of subsets we call the *open sets* such that (1) the empty set and X are both open (2) finite intersections of open sets are open and (3) all unions of open sets are open.

Definition 2.11 A set X together with a topology is a **topological space**.

Definition 2.12 A **continuous** function between topological spaces is a function with the property that the inverse image of every open set is open.

Definition 2.13 A **homeomorphism** is a bijective continuous function between topological spaces with a continuous inverse.

Definition 2.14 A function $f : U \subseteq \mathbb{C} \to \mathbb{C}$ is **holomorphic** on the open set U if the function f is complex differentiable on every point p in U. Differentiability of a complex function is the same definition as for real functions, just with complex values instead of real values. If $U = \mathbb{C}$, then the function is sometimes called an **entire** function. The definition of holomorphic functions is very powerful as, unlike with real functions, being differentiable in the complex plane automatically guarantees the function is infinitely differentiable, and that the function is equal to its Taylor expansion (this last idea is called **analytic**).

Definition 2.15 A **Hausdorff** space is a topological space X in which we can separate all pairs of points by open neighborhoods, i.e., given any two distinct elements $x, y \in X$, there exist neighborhoods U of x and V of y so that $U \cap V = \emptyset$.

Definition 2.16 A topological space is **second countable** if there is a countable base for the topology. A **base** for a topology is a collection of open sets, so that each open set in the topology can be written as a union of the sets in the base.

Definition 2.17 A topological space is **connected** if it cannot be written as the union of two disjoint nonempty open sets.

Definition 2.18 Given a topological space X, a **partition** P of X is a collection of subsets of X, so that each element of X is in one, and only one, of the subsets. With the map $\pi : X \to P$ sending x to the subset of the partition P in which x lives, the set P inherits the topology of X by defining the open sets in P to be subsets U of P so that $\pi^{-1}(P)$ is open in X. In this paper, our P is X/G. We call this inherited topology the **quotient topology**.

2.9 Further Reading

My first introduction to Riemann surfaces was the book [17]. For my own research, I regularly refer to [7]. It has many important results in it, but I would not recommend it as a first introduction to the field. In preparing this paper, I also read through [26]. Group actions are usually not covered extensively in undergraduate level abstract algebra books. If you are ready to read more about them, I recommend Section 1.7 and Chapter 4 in [5]. A reasonable introduction to algebraic geometry for undergraduates is [23]. Do not discount wikipedia.org as a reference for mathematical topics, although the mathematical entries can be dense.

Acknowledgments All figures were drawn by Joe Mileti, who also wore the hat of alpha reader for this paper, as well as being my lodestar in mathematics and in life. Thank you to Deanna Haunsperger and Stephen Kennedy for dreaming up, and pouring so much energy into, the deeply impactful Carleton SMP. I am very grateful to Grinnell College for a sabbatical leave (and funding through a Frank and Roberta Furbush Scholarship) which afforded me the time to write this piece.

Thank you to the anonymous reviewers who gave me insightful comments on places where the exposition could be improved upon. It is not usual to acknowledge ones pet, but our cat Cayley (named after the mathematician, of course) spent the last days of her too short life curled up on my lap as I finished this paper.

References

1. Armstrong, M.A.: Basic topology. Undergraduate Texts in Mathematics. Springer, New York (1983). Corrected reprint of the 1979 original
2. Breuer, T.: Characters and automorphism groups of compact Riemann surfaces. London Mathematical Society Lecture Note Series, vol. 280. Cambridge University Press, Cambridge (2000)
3. Broughton, S.A., Paulhus, J., Wootton, A.: Future directions in automorphisms of surfaces, graphs, and other related topics. In: Automorphisms of Riemann Surfaces, Subgroups of Mapping Class Groups and Related Topics. Contemporary Mathematics, vol. 776, pp. 37–67. American Mathematical Society, Providence (2022)
4. Cox, D., Little, J., O'Shea, D.: Ideals, varieties, and algorithms, 3rd edn. Undergraduate Texts in Mathematics. Springer, New York (2007)
5. Dummit, D.S., Foote, R.M.: Abstract Algebra, 3rd edn. John Wiley & Sons, Inc., Hoboken (2004)
6. Ekedahl, T., Serre, J.P.: Exemples de courbes algébriques à jacobienne complètement décomposable. C. R. Acad. Sci. Paris Sér. I Math. **317**(5), 509–513 (1993)
7. Farkas, H.M., Kra, I.: Riemann surfaces. Graduate Texts in Mathematics, vol. 71, 2nd edn. Springer, New York (1992)
8. Fischer, A., Liu, M., Paulhus, J.: Jacobian varieties of Hurwitz curves with automorphism group PSL(2, q). Involve **9**(4), 639–655 (2016)
9. Fried, M.: Exposition on an arithmetic-group theoretic connection via Riemann's existence theorem. In: The Santa Cruz Conference on Finite Groups (University of California, Santa Cruz, Calif., 1979). Proceedings of Symposia in Pure Mathematics, vol. 37, pp. 571–602. American Mathematical Society, Providence (1980)
10. Griffiths, P., Harris, J.: Principles of Algebraic Geometry. Wiley Classics Library. John Wiley & Sons, Inc., New York (1994). Reprint of the 1978 original
11. Izquierdo, M., Broughton, S.A., Costa, A.F., Rodríguez, R.E. (eds.): Riemann and Klein surfaces, automorphisms, symmetries and moduli spaces. Contemporary Mathematics, vol. 629. American Mathematical Society, Providence (2014)
12. James, G., Liebeck, M.: Representations and Characters of Groups, 2nd edn. Cambridge University Press, New York (2001)
13. Jones, G.A., Singerman, D.: Complex Functions. Cambridge University Press, Cambridge (1987). An algebraic and geometric viewpoint
14. Lee, J.M.: Introduction to topological manifolds. Graduate Texts in Mathematics, vol. 202, 2nd edn. Springer, New York (2011)
15. L'Engle, M.: A Wrinkle in Time. Dell Publishing Company, Co., New York (1962)
16. Milne, J.S.: Jacobian varieties. In: Arithmetic Geometry (Storrs, Conn., 1984), pp. 167–212. Springer, New York (1986)
17. Miranda, R.: Algebraic curves and Riemann surfaces. Graduate Studies in Mathematics, vol. 5. American Mathematical Society, Providence (1995)
18. Munkres, J.R.: Topology, 2nd edn. Prentice Hall, Inc., Upper Saddle River (2000)
19. Needham, T.: Visual Complex Analysis. The Clarendon Press, Oxford University Press, New York (1997)
20. Paulhus, J.: Decomposing Jacobians of curves with extra automorphisms. Acta Arith. **132**(3), 231–244 (2008)

21. Paulhus, J.: Elliptic factors in Jacobians of hyperelliptic curves with certain automorphism groups. In: ANTS X—Proceedings of the Tenth Algorithmic Number Theory Symposium. Open Book Ser., vol. 1, pp. 487–505. Mathematical Sciences Publishers, Berkeley (2013)
22. Paulhus, J., Rojas, A.M.: Completely decomposable Jacobian varieties in new genera. Exp. Math. **26**(4), 430–445 (2017)
23. Reid, M.: Undergraduate algebraic geometry. London Mathematical Society Student Texts, vol. 12. Cambridge University Press, Cambridge (1988)
24. Serre, J.P.: Géométrie algébrique et géométrie analytique. Ann. Inst. Fourier, Grenoble **6**, 1–42 (1955–1956)
25. Silverman, J.H., Tate, J.T.: Rational points on elliptic curves, 2nd edn. Undergraduate Texts in Mathematics. Springer, Berlin (2015)
26. Teleman, C.: Cambridge university lecture notes: Riemann surfaces (2003). https://math.berkeley.edu/~teleman/math/Riemann.pdf. Last visited on 03/13/2018
27. Volklein, H.: Groups as Galois groups: an introduction. Cambridge Studies in Advanced Mathematics, vol. 53. Cambridge University Press, Cambridge (1996)

Chapter 3
Challenges in Using Data for Public Policy Decisions

Laurel Paget-Seekins

Abstract Using data to inform public policy decision is a challenge given the multiple, and sometimes competing, goals for public agencies. Public transit agencies often have large amounts of data, but the data are not always accurate, and there are problems to solve where no data exists. Informing policy outcomes requires identifying the right problem, following the data to find the right research questions, accepting some level of uncertainty, and determining the right way to visualize the results based on the audience. *This paper assumes a limited understanding of statistics and basic knowledge about public transit systems.*[1]

Keywords Public policy · Automated data · Data science · Public transit · Qualitative data · Data cleaning · Spatial analysis · Data visualization · Survey data · Transportation · Data-driven decision-making

3.1 Biography

A lot of my career has been about taking opportunities even if I had no idea where they would lead. I had no idea as an undergraduate student that someday I would be running a data analytics department in a public transit agency. In fact where I started is very far from where I am now, but the path between the two places is both a bit random and entirely logical. Each step has built on the previous ones. Two themes of my story are finding intersections between my interests and building interdisciplinary skills.

I attended Oberlin College because of its progressive history, and raised in a rural area, I wanted to go to college in a small town. I had a double major- mathematics

[1] This paper reflects the author's work as of 2019 when she worked at the MBTA.

L. Paget-Seekins (✉)
Public Advocates, San Francisco, CA, USA

A. Brisbin et al. (eds.), *Research Connections*, Association for Women in Mathematics Series 36, https://doi.org/10.1007/978-3-031-81217-0_3

and a self-designed major I called Liberation Struggles Studies (a mix of history, politics, religion, and economics).

I chose pure math for the beauty; I refused to take any applied classes. I loved that there were right answers, that you could prove something and it was true. The sense of certainty provided a balance to my other major, where I took a lot of classes about social movements to improve the human condition. There were no right answers and no certainty that we could find them.

I attended Carleton College's Summer Mathematics Program between my sophomore and junior year. I can't tell you I remember any of the details of the classes taught that summer, but I do remember the people: Deanna, the professors, and my peers. The program inspired me to do a Research Experience for Undergraduates the next summer. I ended up in a program at SUNY Potsdam studying at graph theory (which, spoiler alert, turns out to have similarities with public transit network design).

The experience of math research was a challenge, because there were no clear steps to take. I didn't know if there was a right answer to find. I also realized I didn't know why any particular problem was worth solving more than another.

When we presented our work from that summer during a poster session at the Joint Mathematics Meetings, the only person who came by and said they were working on similar research worked at the Department of Defense. At that point, I realized I had run into the limits of my strict pure math policy. Without understanding the applications, I didn't know if my results would contribute to the world I wanted to live in or potentially cause more harm. I didn't know which problems to focus on solving.

After undergrad, I did a lot of work in social movements working on social equity, environment, and energy issues. Rather randomly I ended up moving to Atlanta, Georgia, without a car and in the process I realized that public transit is in the intersection of the things I was passionate about. I decided to attend the Georgia Institute of Technology (GA Tech) to get a master's in city and regional planning focusing on transportation. Soon after I arrived, I realized I could put my quantitative skills to use and added a second degree in transport engineering.

For my capstone project, I worked with a community organization of transit riders to develop a plan for public transit in Atlanta from the perspective of people who are dependent on public transit. My goal was to share the quantitative and technical skills I learned in graduate school while learning from the experiences of the people who ride transit every day.

My advisor at GA Tech offered to let me stay on and do a PhD in Civil Engineering. I used it as an opportunity to continue my work with community organizations and hone my research design skills. For my dissertation topic, I returned to the premise that public transit is the intersection of social equity, environment, and energy issues. While working with transit riders in Atlanta, I realized that service designed to get people out of their cars to improve the environment and reduce greenhouse gas emissions didn't always improve access for people without cars. So my research question focused on what network design and operational characteristics of public transit serve both environmental and equity goals. To measure to this, I used nonwork trip usage and nonmotorized access to transit as indicator variables.

I collected a dataset on 17 transit agencies around the USA and used nonparametric statistics (which do not assume a certain distribution) to measure the correlations with my two dependent variables. I also did three qualitative case studies involving interviews with decision-makers in three regions. This mixed research methods approach allowed me to consider how to interpret multiple different types of data at the same time.

When I was finishing my Ph.D., one of my professors sent out an email about a post-doctorial fellowship in mobility culture in megacities at the Technical University of Munich in Germany. I wasn't exactly sure what mobility culture meant, but I applied anyway. I spent 6 months in Germany working with an international and interdisciplinary group of fellows thinking about the different forces that shape transportation in large cities. I also learned about high-quality transit riding trains around Europe.

While in Germany whenever I tried to say something in German, it came out in Spanish, the language I studied in high school. I decided to move to South America, because a lot of advancement in public transit is coming from cities there. One of my colleagues in Munich saw a posting for a job at an international research center on Bus Rapid Transit (BRT) in Santiago, Chile, and forwarded it to me. So I applied.

I ended up spending 2 years in Santiago, Chile, based at a transport engineering department at the Pontificia Universidad Católica de Chile. Most of my research centered on how governments in developing cities are trying to regulate public transit in order to control the problems caused by bus drivers competing on the street to pick up passengers. This informal provision of public transport requires limited public investment but is unsafe, causes congestion on busy streets, and is high in emissions. BRT not only prioritizes bus riders for space on public streets by creating bus only lanes but also provides a means for governments to intervene and formalize the public transport sector.

The research center was multidisciplinary, and one of my bigger projects was editing a book that brought together the work from all parts of the center together to tell a single story. To do this, I drew on my interdisciplinary background and experience finding the intersections of many issues.

After 2 years, I decided I wanted to come back to the USA, and I chose Boston because I was working with researchers from the Massachusetts Institute of Technology (MIT), knew the head of the Massachusetts Bay Transportation Authority (MBTA), and I had friends there. The MBTA offered me a job as the Director of Research and Analysis.

My title has changed over time, but essentially I lead the data analytics department at the MBTA. My team is situated in the agency between the leadership and the operating divisions. We collect and analyze data, large and small, to guide policy decisions and to help our operating divisions improve service. We develop and track performance measures and then do research, run pilot projects, and prototype tools to help improve performance. We also use spatial data to look at how people use and our network and think about how to improve the access we provide.

One of my key roles in this position is acting as a translator between policy-makers with ideas and the operating staff who understand the details of how the systems work. Again my interdisciplinary background is critical to being able to understand the policy ideas and the implementation details. I also translate between academia and government by figuring out how to implement research design in practice. Finally, I have to translate the results of our analyses to nonmathematical audiences in public presentations to our Board of Directors and to the media.

Each part of my journey from undergrad to my current position builds on the skills I developed at previous stages. On this journey, I went from picking a college based on it being in a small town and refusing to study applied math to using applied math to run the infrastructure that makes large cities possible. I have embraced the messiness of applied math and realized that it is part of the solution for making progress on the social problems I wanted to solve as an undergrad. Along the way I have drawn on my interdisciplinary background to translate between different audiences and used intersections to focus my work.

3.2 Introduction

With increasing use of digital technology, more and more data are available. The field of data science develops tools and techniques to harness all of this data to inform decisions. Data-driven decision-making is common phrase for leaders in business and government. While this sounds like a good idea, it is hard in practice. In particular, the use of data in public policy is difficult, because there are always multiple competing goals and a mandate to serve everyone. Often there is no single outcome to optimize, method to use to analyze the data, or way to visualize the results.

A key challenge is figuring out how to use data to solve problems and not just support already reached conclusions or politically expedient answers. Often it isn't intentional; people ask the questions they think they know the answer to, instead of taking the time to follow the data (down a lot of potentially fruitless paths) in order to generate new insights.

There are multiple components to following the data. First, data scientists have to find and make sense of the data available and consider the data not available. Second, they have to ask the right research questions and try multiple analyses. And finally they need to be able to visualize and explain the results to multiple types of audiences to get policy decisions made.

This paper draws on my experience leading a data analytics team inside the Massachusetts Bay Transportation Authority (MBTA). Public transit is a particularly illustrative example, because there is a lot of data and many problems to solve and policy decisions to make. Since public transit has to serve everyone and requires public subsidy, it needs to be equitable and efficient. In addition, public transit exists in both time and space, making longitudinal and spatial analyses possible.

3.3 Public Policy Problem-Solving

Since public policy is about making decisions that are in the public good, most analysis is fundamentally about trade-offs. It is hard to optimize for any single outcome as different groups of people prefer different outcomes. Since public policy is about people, not numbers, we cannot just consider the median or average. The role of government is to serve all people, especially people at the margins.

The majority of time and resources are often spent on the edge cases, because they are the hardest to serve. We build our infrastructure for the peak hour of demand and to withstand the worst weather day. We also build it to serve the minority of people in wheeled mobility devices. It means that many of our analyses are answering the questions who will be impacted by a decision, and by how much.

For example, the MBTA ran a pilot program to determine if providing reduced fares to low-income youth would improve their access to opportunity. Participants in the pilot allowed us to look at their usage of the transit system for 30 days before they received a reduced fare card, and then we monitored their usage afterward. They also answered surveys each month about the types of trips they were taking. This pilot project found with automated data that participants increased the total number of trips they took on the MBTA by 67%, and using survey data found that 14% of trips wouldn't have been made at all without the reduced fare pass. (For more details, see [4].)

The results of this research project convinced the MBTA Board of Directors to create a permanent program to provide these discounts. The program only impacts a very small percentage of the overall MBTA ridership and requires resources to operate but was decided to be worthwhile due to its benefits.

The easiest course of action is to identify problems when we have data to look through. But in the case of the Youth Pass, we had to create a yearlong research pilot in order to collect the data necessary to measure the impact of the program. Just because we don't have data about something, doesn't mean there isn't a problem to be solved. The converse is also true; the existence of data about something shouldn't prioritize it as a problem to be solved.

The need to collect the Youth Pass data came from youth and transit advocates putting political pressure on the MBTA for years. Usually, our data team gets questions from agency leadership, our operating departments, and often from our own curiosity. Once we have a question, it can take several hours to answer or several years.

The kinds of questions we are using data to answer ranges from general to very specific questions. For example, our leadership asked, "how much fare revenue does the MBTA lose by allowing people to board at the rear doors of crowded light rail trains above ground in order to speed up operations?" In order to answer this, we created a sampling plan by looking at the ridership data by time of day, went out and rode trains to collect data, and the scaled it up. We presented the results to agency leadership who used it to reaffirm that the operating time savings was worth the revenue loss.

Often people come to us asking for data and our response is usually "what problem are you trying to solve"? We have realized that we need to understand the problem before we know what data would be the most useful.

3.4 Collecting and Cleaning Data

Before we can solve any problems, we need to understand the data sources, so we can try to turn them into information and hopefully knowledge. The well-used data information knowledge wisdom pyramid illustrates this hierarchy (there are multiple sources for this graphic, see [3]) (Fig. 3.1).

Data can come from many sources, but often the data is a by-product of the original intent of that source. For example, many transit agencies use their automated fare collection system to measure ridership and other aspects of how people use their system. However, the primary intent of a fare collection system is to collect fares, not to count people. So whether due to people evading paying or being free due to the rules of the transit agency, the system won't get a complete count of everyone.

Figuring out the limitations of the dataset is critical to knowing what types of bias it might introduce into an analysis. For most problems, there is not a perfect source of data to analyze. Clearly understanding the problem, and why you are trying to solve it, is the first step to determining which of your available datasets are best suited or if you need to gather new data.

At the MBTA, we have many automated sources of data, and we do data collection to supplement it. Our main automatically generated datasets are the fare collection system and the vehicle location data. These systems tell us where people enter the system, how they pay, and whether the vehicles are on time. While sometimes our automated systems go down and we lose data, in general, these datasets are not a sample but the entire population of data points. This means they have millions of records and that are updated daily.

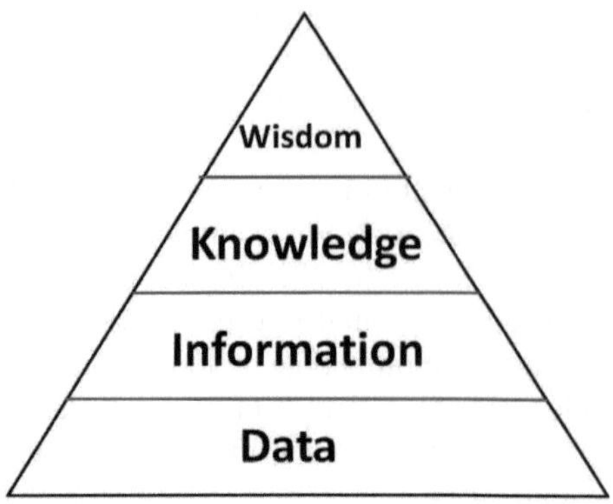

Fig. 3.1 The data information knowledge wisdom pyramid

But they don't tell us who our passengers are and what they think about the service we provide. To gather this information, we have to survey a sample of our riders. We conduct multiple types of surveys to better understand our passengers, their travel needs, and the impact of changes we make. In order to be able to draw conclusions about our entire population of passengers, we have to design a sampling strategy based on the level of confidence and precision we need and the resources we have available to conduct the survey.

Once we have a dataset, whether it is from a survey or automated source, the first step is cleaning it. When we use large automated sets, we have to account for times when the data wasn't collected, either due to technological failures or unique situations. For example, the internal clocks on our fareboxes drift by a few seconds, so in order to match the fare collection data to the vehicle location data, we have to run a process to account for the drift. Occasionally, we open the faregates due to large crowds in stations after a special event for safety reasons, but then we aren't collecting any data of ridership counts.

Sometimes we find mysteries in the data, and it takes us months of investigating to figure out why. For example, low ridership counts in late June and early July one year turned out to be due to a software upgrade to the fareboxes on vehicles as part of a fare increase. (Often these data mysteries are identified further into the process when we get results that don't make sense.)

We also clean our survey data. This includes reading and sorting all open text fields but also includes checking for outliers and the distribution of variables. Most important, we have to decide how to weight the data, so we can make statements about the population we want to talk about. We usually weight based on ridership on each of our types of services (commuter rail, subway, bus, etc.), because we have found that what service someone is riding is a defining characteristic of how they perceive the service, and because people from different types of services respond to surveys at different rates.

We know that all of our data are wrong to some extent. One of our main jobs is determining the level of wrongness and whether, based on the problem we are trying to solve, we are comfortable with that amount of accuracy. In order to do this, we need to understand the context under which the data was collected. For example, can we assume the error is randomly distributed or is there relevant bias and what direction is it in?

While our data team understands the nuances of the datasets and the sources of error, it is difficult to stress the nuances enough with our multiple audiences. People want to believe numbers and rarely want to read scores of footnotes on each result. We have to make a judgment call about what data are accurate enough to use for each problem. The next step is analyzing the data to turn it into information.

3.5 Turning Data into Information

After determining the problem we are trying to solve and the datasets available, we have to organize data for analysis. Some of the initial considerations are what level of data aggregation, over what time-frame, and what units of geography or service type. There is no single way to answer these questions and often require doing the analysis multiple ways.

For example, how should we compare ridership over time? We have to consider how many weekdays and how many holidays in a year or a month that some months have higher ridership than others, and the impact of service or fare changes. In addition, we need to account for changes in data collection methodologies. In order to account for all of these factors, we conduct multiple comparisons. We normally look at median ridership to account for outlier days or months, compare month over month to account for seasonality, and look at each type of service (bus, subway, commuter rail) and day type (weekday, Saturday, and Sunday) separately. We look at recent trends (last 2 years) and long-term trends (last 10 years) factoring in internal and external changes.

For spatial data, we have to consider what geographic units to use. This is dependent on the data sources available. We often use US Census Bureau data for population and jobs attached to census tracts. We want to match ridership at our bus stops and stations to census data, but often our stops are on the border between two census tracts. So either we have to assign our ridership between census tracts or assign census data to different geographies.

One of our solutions to this problem of combining data from different datasets was to create our own unit of geography using k-means clustering (this is a method to divide data points into clusters around a mean based on set parameters). We used spatial software to create geographies that contained our stops where each geographic unit had roughly the same number of riders boarding each day. So units in the center of Boston near downtown stations are quite small and units in the suburbs with less ridership are quite large.

Once we have the data aggregated into units of analysis, the next step is what analysis to do. This depends on the problem and can range in complexity. Often it is as basic as creating graphs and looking at trends or variability. Other times our analysis involves complicated models that can take multiple years of work to develop.

For survey data, we measure the change from month to month in average satisfaction levels. We also use linear regression modeling to look for significant variables that explain change in our top-line satisfaction questions. Since one of our survey instruments is a monthly panel survey where the same people answer questions each month, we also are able to do longitudinal analysis. We can look at how individuals' satisfaction changes over time based on what services they use, demographics, or their past experiences. A key consideration is variability. On which variables are people more likely to change their answers over time? And are some people more variable than others?

Working with researchers at MIT, we have developed an origin destination inference model. Our fare collection system only records where someone boards our system using our smart fare card or ticket. But we want to know where people are going, so we can better plan our service. Looking at the patterns of the fare cards and tickets over the course of a day, we infer where people are getting off. (For more details, see [1]).

This is a complicated model that we run after we have all of the data from each quarter of the year. It is stored in a relational database, and we use SQL queries to pull data from it. We also build data visualization tools and other analyses on top of it.

One of problem we use this model to analyze is how to address crowding on buses. As part of a process to develop standards for quality and quantity of service we provide, we set a standard for crowding on bus routes based on the percent of passenger minutes spent in crowded conditions. Crowding on buses exists in both time and space, so we visualize it for each street segment and 30-minute time periods on an average day.

There are multiple possible ways to fix bus crowding. Just adding buses to the schedule won't help if the crowding is caused by buses bunching up or trips not running due to a lack of buses or operators. So with MIT, we built a model of the causes of bus crowding on each bus route to help us identify what tools to use.

The model is based on simulating the demand in four cases. If the passenger demand in a time period is even across all days during an analysis period and there is still crowding based on the supply provided by all of the scheduled trips, then we need to add more service. If we don't assume even demand across days, but allow it to vary and there is crowding above the first case, then it is likely due to variability in demand (e.g., some weekdays are busier than others). In this case, we might need to not schedule for the average load, but a higher percentile of the load. The next case looks at the crowding on the trips actually run, not the ones scheduled. If there is crowding above the other two cases, it could be the result of not all of the scheduled trips running (trips can be canceled if not enough buses or operators are available). Finally, we look at how on time the service was operated compared to the schedule to see the impact of unreliability. (For more details, see [2].)

All of these analyses require different analytical skills and software programs, but the central theme is figuring out what questions to ask. The analysis process isn't linear. Sometimes partway through an analysis you realize the data is wrong, you need a different level of aggregation, or that your initial question was the wrong one. You can keep analyzing forever, but at some point, you have to feel confident enough about results to present them, knowing that improvements can still be made.

3.6 Converting Information into Knowledge

The final step is figuring out how to present the results depending on the audience to inform the decision that needs to be made, or how can we turn information into knowledge. In the case of the MBTA, our analysis may be for operating departments, internal leadership, or public presentation to our Board of Directors, the public, and the media.

We have to figure out what level of detail is needed and when we need to mention important caveats and considerations about the data. None of our audiences is particularly interested in whether our regression analyses pass normality assumptions. But often we need to explain what conclusions can and cannot be drawn from our results.

In some situations, we want to present the same results to multiple audiences. We built an online interactive performance dashboard for the public but used a layering approach. So the top layer is accessible to the most general audience and then people interested in more technical details can drill deeper. (For more information, www.mbtabackontrack.com.)

A key challenge is how to communicate that results aren't static and that it is possible with additional or better quality data our results will change. This is difficult when we are presenting results publicly. Uncertainty is a hard concept in general, but especially for the public sector as people rarely want their government to admit they don't know the answer.

One solution we use to explain data caveats and address uncertainty is a data blog (https://www.massdottracker.com/datablog/). The data blog allows us to explain at the level of detail we want: the analysis we do, the assumptions we made, the problems with the data, and the reasons for any mistakes. As we improve data quality we can explain the methodological breaks this creates and how this limits trend analysis.

It is satisfying when months or years of work we have done is publicly presented and used to inform a decision. But often it is boiled down into a few charts and tables. The data blog also gives us a place to tell the story behind that chart and the work that went into it.

When preparing for public presentations, we try to remember that good data analysis is storytelling. We want to remind the decision-makers why this is a problem we are trying to solve and how it impacts people. Sometimes it is possible to aggregate data past the point of people and forget the initial reason for the analysis. This is why we often pair our quantitative analysis with qualitative data we collect such as interviews with passengers, comments we receive on surveys, and photos we take.

3.7 Conclusions

Our data team has conducted numerous analyses that have impacted decision-making at the MBTA. Our results cannot always tell the decision-makers the optimal outcome for everyone impacted, but we can provide context and guidance. Due to our public mission, we make sure that in the aggregation of our analysis, we don't lose track of the people most impacted by the decision.

Our analytic process is guided by the data to wisdom pyramid. First, we ask what problem we trying to solve. Then we look for the right data and if we don't have it, we figure out how to collect it. A large amount of time is spent understanding the context of how the data was gathered and cleaning it. We think about how to aggregate it and the best analysis techniques. After starting analysis, we recalibrate as we go. The analysis process helps us determine the level of uncertainty we are comfortable with and accept that our results aren't static. Finally, we decide how to visualize and present our results given our audience and the initial problem we are trying to solve.

The most satisfying results are those that change (or at least challenge) an existing narrative, where the data tells a different story than what we thought was true at the beginning. In order to get these results, we have to have an open mind and ask the right questions. It requires taking the risk of going down paths that will generate not interesting results. Often the things that are not statistically significant are just as important as the things that are.

In order to produce these results, we have to have time and freedom to explore. Much of the analysis we conduct is not requested by decision-makers, but self-driven by our intellectual curiosity and desire to improve public transit. We ask questions and then we try to answer them. Sometimes it takes us years to do, which is a luxury in a government agency with limited resources. Part of what helps us is our robust intern program. We hire undergraduate and graduate interns, from many academic backgrounds, to help us explore our datasets, solve data mysteries, and visualize our results.

This work can only be done by a diverse team; it requires a variety of skills and experiences. This includes statistics, spatial analysis using GIS, database management, front-end web development, graphic design and data visualization, survey design, policy analysis, and process improvement. Some team members studied transportation planning, engineering, or planning, but we all share a commitment to improving our urban region for the public good. We have many ideas for future research projects; the only limit is the time we have to do them.

References

1. Sánchez-Martínez, G.E.: Inference of public transportation trip destinations by using fare transaction and vehicle location data: dynamic programming approach. Transp. Res. Rec. **2652**(1), 1–7 (2017)
2. Sánchez-Martínez, G.E., Paget-Seekins, L., Southwick, C.W., Attanucci, J.P.: Bus load inference and crowding performance evaluation through disaggregate analysis of fare transaction, vehicle location, and passenger count data. Transp. Res. Rec. **2672**(8), 464–474 (2018)
3. Sharma, N.: The origin of data information knowledge wisdom (DIKW) hierarchy. Preuzeto **25**, 2021 (2008)
4. Thistle, I., Paget-Seekins, L.: The youth pass: a study of the conflicts between ideal and practical research design in a pilot social program. Transp. Res. Rec. **2652**(1), 116–123 (2017)

Chapter 4
A Taste of Discretized Differential Geometry: Communicating Mathematics with Conceptual Metaphor

Melinda Lanius

M. Lanius (✉)
Auburn University, Auburn, AL, USA
e-mail: melinda.lanius@auburn.edu

A. Brisbin et al. (eds.), *Research Connections*, Association for Women in Mathematics Series 36, https://doi.org/10.1007/978-3-031-81217-0_4

Abstract We will explore the art of discretizing, a way of building something new from a familiar and well-understood mathematical object: For instance, the real number line can be "discretized" into the integers. Our discussion will center on mathematical objects drawn from my area of specialization: differential geometry. This area of math uses linear algebra and calculus to understand the geometry of objects like a ball or a cone. Our meandering discussion will include puzzle games, a ton of pictures, and questions for you to explore. We assume that the reader has had a full year of calculus (Calc. I and II) and possibly multivariable calculus (oft times called Calc. III) as well. Some examples and remarks will involve graph theory and topology; but familiarity with these subjects should not be necessary to follow the main idea.

Keywords Conceptual metaphor · Discrete differential geometry · Discrete · Continuous · Geometry · Topology · Discretization · Tiling · Torus · Research in undergraduate mathematics education

4.1 Biography/Career Trajectory

My research lies in Math-Discipline-Based Education Research (Math-DBER), an interdisciplinary research area that utilizes the fields of Mathematics, Data Science, and Education to investigate the teaching and learning of mathematics at the undergraduate or graduate level. While Math-DBER is informed by more general research on learning and cognition, it is distinct in that the research questions are grounded in the priorities, knowledge, cultures, philosophies, and histories of the mathematics community.

While my current research area focuses on education, my undergraduate and doctorate degrees are in Mathematics; I graduated from Wellesley College in 2012 and the University of Illinois at Urbana-Champaign in 2018. After my 2018 graduation, I moved to the University of Arizona for a 3-year post-doc, a type of temporary stepping-stone position that has more expectations than being a graduate student, but fewer than being a full-time faculty member. In the second year of my post-doc, I realized that I was fascinated by *math as a verb* rather than *math as a noun*. In other words, while I found the subject of math to be cool, I was captivated by the different ways people learn, create, and communicate mathematics. DBER, as a research area, not only enables me to leverage my Math PhD to engage in education research, it actually values and utilizes the skills I developed as a mathematician. Looking back, I am grateful I did a post-doc position immediately after graduation, instead of trying to jump into a permanent faculty position. In 2018, I still wasn't sure what I wanted to do with my career and the post-doc gave me the space to explore and find the right option for me. Now, I am an Assistant Professor at Auburn University – where I was hired to specifically conduct Math-DBER in a Mathematics and Statistics Department – and I feel certain and at home in my research area.

I wrote the following chapter in that 2018–2019 transitional period, where I identified as a research mathematician but didn't feel quite at home in my area (geometry). From my position today, I can identify components in my writing that later informed my DBER research. In particular, my exposition heavily utilizes *conceptual metaphor*. Conceptual metaphor is a way that people can understand challenging abstract concepts using metaphor. In the following chapter, you will see me use two types of metaphor: *grounding metaphor* and *linking metaphor*. Grounding metaphors allow people to understand abstract concepts by projecting their everyday experiences onto those concepts; linking metaphors, on the other hand, enable people to understand an abstract concept by linking it to another abstract concept that they already understand. If, after reading this chapter, you're interested in how I use conceptual metaphor in my current research, I recommend my papers [10] and [11]. To read more about conceptual metaphor generally, I recommend two books that introduce the idea: [6] and [7].

4.2 The Philosophy of Discretization

How do computers "do" math? Does my trusty calculator have an inexhaustible knowledge of all the values of cosine for any real number x? Well, probably not. You can imagine that there is not enough space to fit all of this information in a computer the size of your hand. Instead, when we ask a calculator to find the value of $\cos(1)$, it uses a type of polynomial expansion – called a Taylor series – that you may have seen in a calculus course. You might remember that the Taylor series expansion of cosine of x is given by the infinite sum

$$\cos(x) = \sum_{n=0}^{\infty}(-1)^n \frac{x^{2n}}{(2n)!} = 1 - \frac{x^2}{2!} + \frac{x^4}{4!} - \frac{x^6}{6!} + \frac{x^8}{8!} - \dots$$

Let's consider a calculator that provides the value of $\cos(1)$ up to two decimal places. We will look at what happens when we truncate the above Taylor series to finitely many terms – or shorten the series by cutting off the tail of the sum past some natural number N.

If we truncate at $N = 1$, then we have approximated that $\cos(1) \approx 1$.. However this is not accurate to two decimal places, because the error or difference between 1 and $\cos(1)$ is approximately 0.4597. So we should consider truncating our series at a later term. By analyzing the table in Fig. 4.1, since the first two decimal places of any truncation remain the same after $N = 2$, we see that any truncation past and including $N = 3$ gives sufficient accuracy for our calculator. In other words, the error or difference between the value of $\sum_{n=0}^{N}(-1)^n \frac{1}{(2n)!}$ and $\cos(1)$ is less than 0.01 for any $N \geq 3$.

$\cos(1) = 0.5403023058\ldots$

N	Approximations of $\cos(1)$
1	$1 = 1$
2	$1 - \frac{1}{2!} = 0.5$
3	$1 - \frac{1}{2!} + \frac{1}{4!} = 0.541\bar{6}$
4	$1 - \frac{1}{2!} + \frac{1}{4!} - \frac{1}{6!} = 0.5402\bar{7}$

Fig. 4.1 Taylor truncations

This idea of truncating the Taylor series, rather than considering the entire sum, is an example of a discretization, because instead of considering an infinite number of terms, we consider a finite number. Although this is not the only way discretizations arise, we will soon see that restricting our attention to a finite part is commonly thought of as a "discretization."

Outline In this paper, we will explore discretizing objects in differential geometry. We will begin in Sect. 4.2.1 by discussing more examples of discretization. Sect. 4.4 will discuss discretization in the field of geometry. As a bridge between these two, Sect. 4.3 will introduce an area called topology. One can think of Sect. 4.3 as a place to define a triangle. That way in Sect. 4.4, if we want to discuss interior angles, we have an object to consider.

4.2.1 Discretization by Example

Discretization is a mathematical approach; one could even call it a mindset. Mathematicians attempt to carry out discretization, for which there is in general no exact recipe, in such a way that we can find analogies between so-called continuous and discrete theory. Rather than leading with a definition of discretization, we will first consider some examples that illustrate the philosophy.

Example 4.1 (Discrete Disco) Dancing is typically a continuous activity. The robot, a style known for its disjoint movements and abrupt stops, is a discretization of dance. □

Example 4.2 (Image Sampling) We discretize the image of Black Forest cake[1] – shown in Fig. 4.2a – by overlaying the image with a 20 by 20 grid. We color each box by the average of the colors appearing directly below. The result in Fig. 4.2b is a discretization, or highly pixelated analogue of the original image. □

[1] The author might have been watching the 2015 series of *The Great British Bake Off* during the production of these notes. Fun fact: The Black Forest cake is called Schwarzwälder Kirschtorte in Germany, its country of origin, and is made up of two layers of chocolate cake with whipped cream and cherries in between.

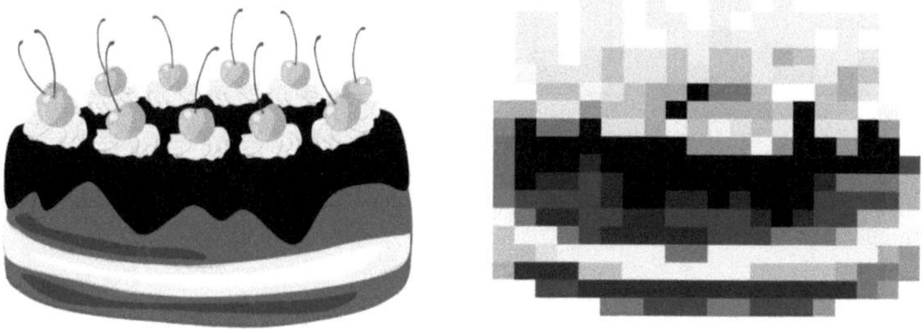

(a) (b)

Fig. 4.2 Images of Black Forest cake. (**a**) Continuous. (**b**) Discretized

Fig. 4.3 Graph of $f(x)$ and discrete analogue a_n

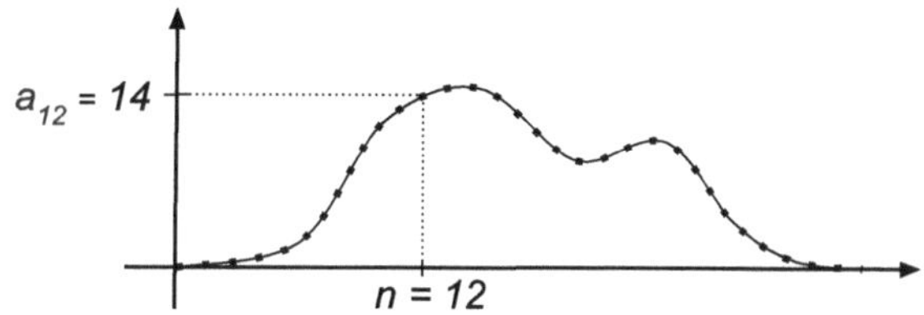

Example 4.3 (Functions & Sequences) Given a real function $f(x)$, we can define a sequence $a_n = f(n)$ for $n = 1, 2, 3, \ldots$. Figure 4.3 compares the graph of some $f(x)$ to the corresponding a_n. Since $f(12) = 14$, the twelfth term of the sequence is 14.

We can consider this process to be a discretization of time itself. In continuous time, between any two points in time, there are infinitely many other points. The variable "time" ranges over the entire real number line. □

Discretization of time mimics the way you take data in a lab. We can't, despite our best intentions, collect data at every possible value of time. Instead, we take a measurement at specific points or moments in time.

Example 4.4 (Integration) Given a real function $f(x)$, the definite integral $\int_a^b f(x)\,dx$ is rigorously defined as the sum of areas of certain rectangles as the number of rectangles increases and the widths of the rectangles become thinner and thinner.

We can view the definite integral geometrically as the area of the region in the xy-plane that is bounded by the graph of f, the x-axis, and the vertical lines $x = a$ and $x = b$, shown in Fig. 4.4a. A discretized version is to approximate this area using a finite number of rectangles. One discretized version of the definite integral comes from the Riemann integral and is the area under the curve approximated using nine vertical rectangles; see Fig. 4.4b. Another discretized version comes from the Lebesgue integral, where we approximate the same area using seven horizontal slabs; see Fig. 4.4c. □

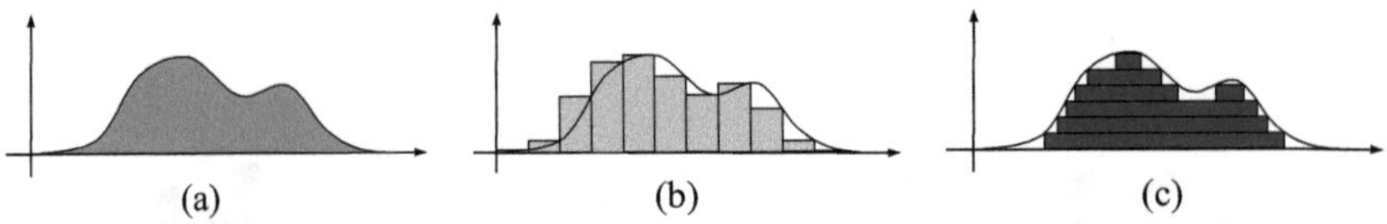

Fig. 4.4 Variety in finding "area under the curve". (**a**) Integral gives area under the curve. (**b**) Riemann integral. (**c**) Lebesgue integral

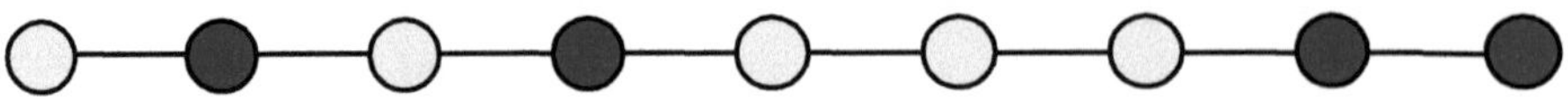

Fig. 4.5 Sperner's lemma

Example 4.5 (Intermediate Value Theorem & Sperner's Lemma) The Intermediate Value Theorem tells us that given a continuous function f defined on the interval $[a, b]$ and a number c with $f(a) < c < f(b)$, there exists at least one x with $f(x) = c$.

One discretization is Sperner's Lemma, due to Emanuel Sperner in 1928 [13]. Consider the path graph in Fig. 4.5. The vertex on the left is colored white, and the right-most vertex is colored black. The vertices in between can be arbitrarily colored either white or black. Sperner's Lemma tells us the color as we travel from one end of the path to the other must switch an odd number of times.[2] □

To conclude, a discretization of a continuous object is a discrete analogue. The act of discretizing is dividing a continuous object into a finite number of discrete elements. Note that this is not a precise mathematical definition like the definition given for a differentiable function or a convergent series in your calculus class. This definition leaves ambiguity or freedom, depending on your outlook. One mathematician can provide a discretization of a mathematical object that is very different from another mathematician's discretization of the same starting object.

4.2.2 *Why Discretize?*

Analogizing An analogy between two disparate areas of mathematics can be very inspiring, leading to great innovations. Further, comparing two results can deepen our understanding of each. For example, the Intermediate Value Theorem inspired Sperner's Lemma. The function f in the Intermediate Value Theorem corresponds to the path graph in Sperner's Lemma. Then the numbers a and b are analogous to

[2] Bernard Bolzano proved a special case of the Intermediate Value Theorem in 1817 [3] and Augustin-Louis Cauchy provided a complete proof in 1821 [4].

the two ends of the path and finding an x so that $f(x) = c$ corresponds to having an odd number of color changes.

Intuition A discrete version can be easier to explain and less technical. These discretizations frequently behave similarly to the continuous example and help us develop intuition for a subject. Building intuition will be our primary motivation when discussing discretized topology.

Computing Discretization is commonly a first step to make a procedure suitable for implementation via computers. Computation motivated the discretizations found in the truncated Taylor series and integration examples. Discretized differential geometry – which we will learn more about in the coming sections – studies discretized geometric spaces. This blossoming field has applications in anatomy, architecture, and the development of computer graphics.

4.2.3 **How** *to Discretize?*

As we mentioned above, discretization is a philosophy rather than an exact process. Computer scientist Keenan Crane and mathematician Max Wardetzky recommend the following steps for discretizating a continuous object [5]:

1. *Write down every description or definition of the continuous object you know.*
   In Example 4.4, we looked at two ways to define the definite integral, either through using vertical rectangles in the Riemann integral or by using horizontal slabs in the Lebesgue integral.
2. *Apply each definition or description to an object in the discrete setting.*
   Following the Riemann approach, our discretized definite integral was the area of nine vertical rectangles. Using the Lebesgue notion of definite integral, our discretization was the area of seven horizontal slabs.
3. *Analyze the advantages and disadvantages among the resulting discretizations.*
   Using vertical rectangles has the advantage of being familiar to more people. Lebesque integration is less commonly taught in Calculus I. In the end though, both of these descritizations seem equally useful.

Commonly two different discretizations of a continuous object behave quite differently. Each may share only some, and different, properties with the continuous theory. It is easy when learning mathematics to think of the subject as monolithic, with a single right answer. But this is hardly the case, and a choice of discretization may simply be a matter of taste.

To illustrate, let's consider pizza: Circle cut and square cut are two possible discretizations of a pie (see Fig. 4.6). Both divide the pizza into eight pieces, but

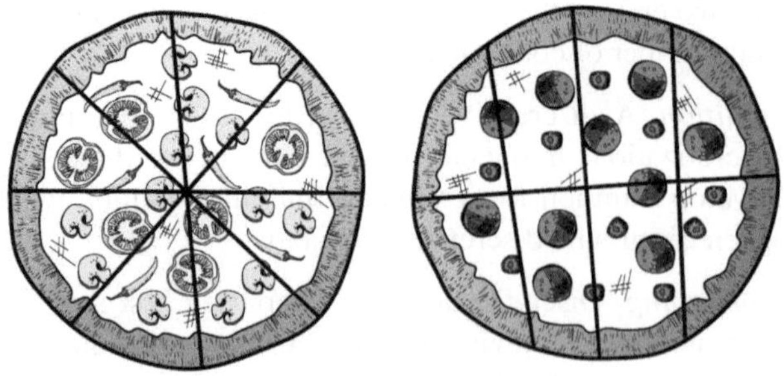

Fig. 4.6 Circle cut vs. square cut

Fig. 4.7 Potential images of a pixelated parrot

people will sometimes surprise you with a very loud and vocal preference for one cut over the other[3].

Exercise 4.1 Consider the image of the parrot in Fig. 4.7. Which of the pixelated images could be a discretization?

Exercise 4.2 Given the discretized function in Fig. 4.8a, which graph in Fig. 4.8b depicts a function with this discretization? Is there more than one possibility?

Exercise 4.3 We have seen some benefits of using a discretized version of a continuous object. Can you think of any disadvantages? In Exercise 4.1, any of the parrots could be a discretization of the original image. All of the continuous functions in Exercise 4.2 have the discretized function as a possible discretization. Is information "lost" going from the continuous object to a discretization?

---

[3] **Stephen:**. The little triangular corner pieces in a square cut pizza are alluring tiny novelty sized slices!

Cindy: You're wrong! It is a disproportionate amount of crust with a mere drop of sauce, a paltry fleck of cheese, and hardly ever contains any toppings.

Stephen: But the square cut maximizes a pizza's shareability! It's easier to cut each piece in half to make more slices!

Cindy: We don't want to *share* pizza, Stephen. We do it because it is the only socially acceptable option.

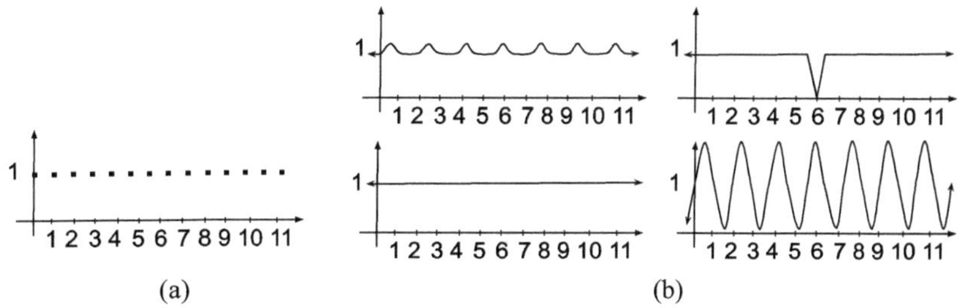

Fig. 4.8 Graphs of functions. (**a**) Discretized function. (**b**) Graphs of continuous functions

4.3 Discretized Topology

Let us put discretization to the side for a moment. In order to discuss discretized geometry, we need to first understand what are called two-dimensional spaces in an area of mathematics called topology.

4.3.1 Topology

Intuitively, topology is Play-Doh[4] geometry. Topologists imagine that physical objects are made from putty. They are interested in the geometric properties and spatial relations that are unaffected by mashing the putty. The ***topology*** of a space can be thought of as the set of properties of it which are unchanged by deforming it.

Example 4.6 A topologist cannot distinguish between a doughnut and a coffee cup. Imagine starting with a Play-Doh doughnut. We can roll the ring into a handle for the mug. To create the cup, we work a piece of clay to one side of the handle and create an indent with our thumb (see Fig. 4.9). We say that the mug and doughnut are **topologically the same**. □

We will focus on a particularly nice type of space called a surface. Surfaces are two-dimensional objects. One familiar example is called the plane $\mathbb{R}^2$, depicted in Fig. 4.10a. You can imagine the plane as a giant flat piece of paper that extends infinitely far in all directions. A space is a ***surface*** if, in any small enough region, it looks like the plane $\mathbb{R}^2$.

The skin of a watermelon and the outside of a doughnut, depicted in Fig. 4.10b, define surfaces. To verify that these are surfaces, we can check that at any point the space looks like a plane. Imagine the perspective of an ant standing on the food.

[4] Play-Doh is a pliable putty, originally marketed as a tool to clean coal dust off wallpaper [14]. It is now more commonly used as a children's toy.

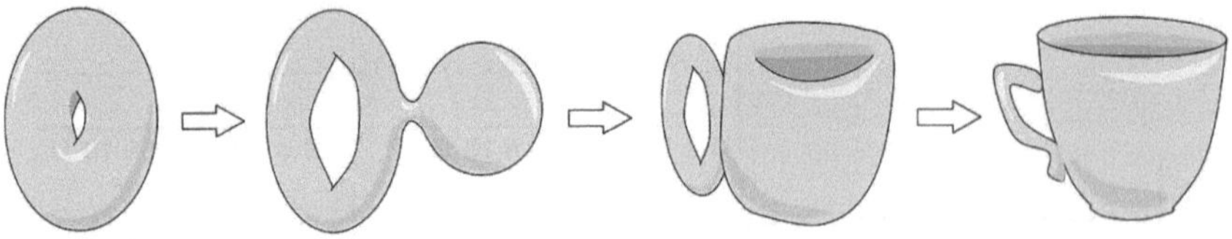

Fig. 4.9 Doughnut transforms into a coffee cup

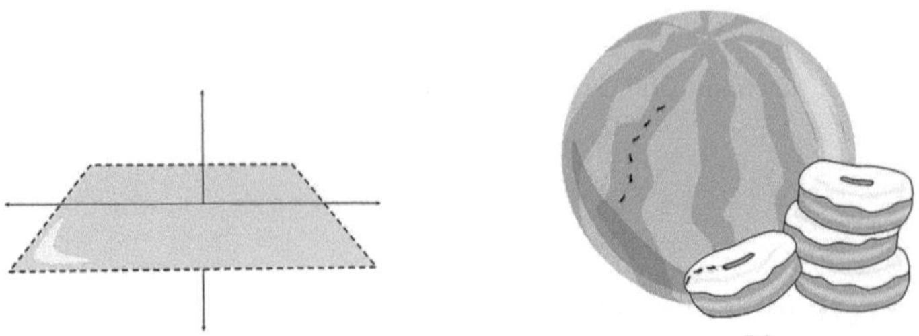

(a) (b)

Fig. 4.10 Surfaces. (**a**) The xy-plane is given by the equation $z = 0$ in Euclidean three space $\mathbb{R}^3$. (**b**) An ant's perspective

(a) (b) (c) □

Fig. 4.11 Exploring what is and is not a surface. (**a**) Torus. (**b**) Hyperboloids. (**c**) Three planes

To the ants, a watermelon or a doughnut looks flat close to where they stand. They can't tell that the space curves or that if they keep walking forward they will return to where they started.

Example 4.7 A torus is shown in Fig. 4.11a. This hollow doughnut is an example of a surface. Next we will consider Fig. 4.11b. The single-sheeted hyperboloid – the outermost part – is a surface. The two sheeted hyperboloid – the two innermost pieces – is also a surface. However, the double cone is not a surface, because the space does not look like a plane at the cone tip. The intersection of the planes $x = 0$, $y = 0$, and $z = 0$ – shown in Fig. 4.11c – is also not a surface. Away from where they intersect, any point in the collection will look like a point in a single plane. However, any point on the lines where they intersect does not. □

4.3.1.1 Compact, Connected, & Orientable

We will study surfaces that are compact, connected, and orientable. To understand the meaning of compact, we must first explore the idea of being closed and bounded.

A space is closed if it contains all of its limit points. To clarify this notion, think of the open interval (0, 1). This is a one-dimensional space where each number between 0 and 1 is considered a point in the space. The numbers $\frac{1}{n}$ for n a counting (natural) number are all contained in this space. However, the number $0 = \lim_{n\to\infty} \frac{1}{n}$ is not contained in the interval (0, 1). Thus, the space (0, 1) does not contain all of its limit points and thus is not closed.

On the other hand, the closed interval [0, 1] does contain all of its limit points. In other words, we cannot find a sequence with values in [0, 1] whose limit is not also in [0, 1].

Bounded means that every point in the surface is within some fixed distance of any other point. In other words, we can hold a copy of the surface in the palm of our hand. Since we can hold a doughnut, the torus is bounded. Since we cannot hold an infinite sheet of paper in our hand, the plane is unbounded. A surface is ***compact*** if it is closed and bounded. The torus is compact. Conversely, the hyperboloids and the plane are not compact.

Next, imagine taking a strip of paper, twisting one end by 180 degrees, and then joining the two ends to form a loop. This is called a Möbius strip, shown in Fig. 4.12. (This is what we call a surface with boundary: any point away from the edge locally looks like the plane. At the boundary edge, the space looks like a half plane – like the edge of a piece of paper.) If you trace your finger along the boundary edge of a Möbius strip, you will find there is only one piece. Additionally, if you trace your finger along the center circle, you will find that the strip only has one side. Intuitively, we should think of orientable surfaces as those that have a distinguished "inside" and "outside." Non-orientable surfaces only have a single side. A surface is ***orientable*** if it does not contain a Möbius strip. A surface is ***non-orientable*** if it does contain a Möbius strip.

Connected means that the surface isn't comprised of multiple disjoint pieces. Every space that we have considered so far, except the two sheeted hyperboloid (comprised of the darkest two pieces) in Fig. 4.11b, is connected. In this paper, we will focus on connected, compact (meaning closed and bounded), orientable surfaces.

Fig. 4.12 Möbius strip

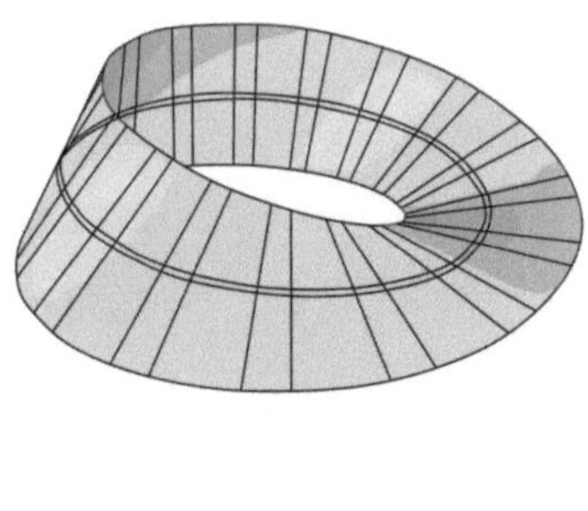

4.3.2 Classification of Compact Oriented Surfaces

A ***classification***, a grouping of structures into classes based on some shared properties, tells us when two structures are the same and allows mathematicians to work with the friendliest model.

Example 4.8 The equations

$$y = x + 1 \text{ and } 2(y - x) - 2 = 0$$

represent the same line, graphed in Fig. 4.13. In practice, we prefer to work with the former expression as the latter clouds our understanding.

In general, lines can be written in the standard model $y = mx + b$, where m is the slope, or steepness of the line, and b is the y-intercept, or how the line crosses the vertical axis. In fact, lines are classified by their value of m and b, meaning two lines are the same if and only if they have the same value of m and b. □

The basic example of a compact oriented surface is the sphere, or hollow ball. The next example is a torus, or hollow doughnut. Every other connected compact oriented surface is specified by how many additional "holes" it has; see Fig. 4.14.

Theorem 4.1 *We say two surfaces are topologically the same when we imagine the two surfaces as made from Play-Doh and can deform one into looking like the other. Any compact connected oriented surface is topologically the same as either the sphere or a torus with k holes, where k is a counting number.*

The theorem says that the outside of a coffee cup is topologically the same as the outside of a doughnut, which itself is topologically the same as a torus. Theorem 4.1

Fig. 4.13 Graph of a line

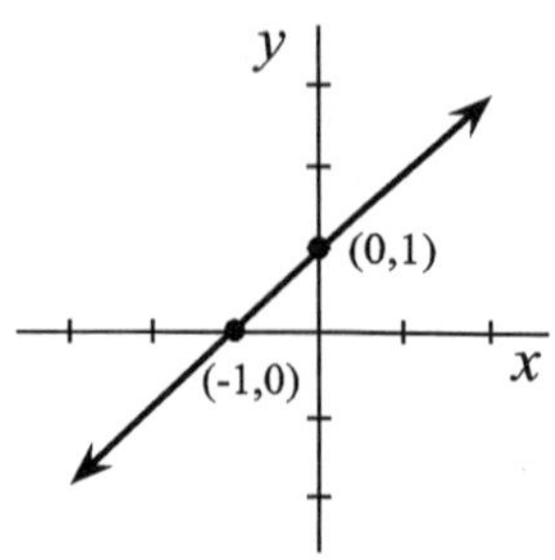

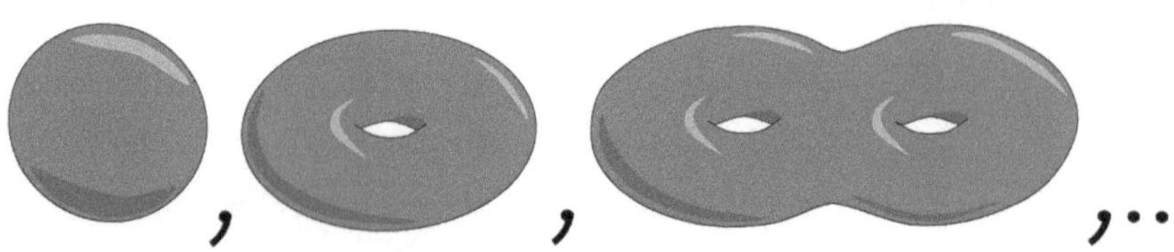

Fig. 4.14 A list of all connected compact oriented surfaces up to deformation

has been known since the late 1800s and has been proven in a variety of ways; the most popular modern proof can be found in Chapter 6 of [1].

4.3.3 *Topology Discretized: Identification Spaces*

It turns out we can construct any compact connected oriented surface by taking a polygon with an even number of sides and gluing certain pairs of edges together. To obtain the sphere, we begin with what we call a 2-gon, a two-edged shape. We label each edge with arrows pointing the same direction, as shown to the left in Fig. 4.15. These arrows tell us that we should glue these edges together to obtain the lemon-like object in the center of Fig. 4.15. This does not appear to be a sphere, since it more closely resembles a football. However, we just treat the space like it is made from Play-Doh and smooth out these pointy bits to obtain the sphere shown to the right.

We can make a torus by gluing together opposite sides of a square (or 4-gon), shown in Fig. 4.16. We label the top and bottom with arrows A pointing in the same direction. We label the right and left edges with arrows B pointing in the same direction. We first glue together the A edges to obtain a cylinder. Next, we bend this cylinder around and glue together the two circles labeled B on either end to obtain the torus.

There are other n-gons with identified edges that will give us a torus. Consider the hexagon with labeled edges given in Fig. 4.17. The labeling of the left sides and the top and bottom are quite similar to that of the square in Fig. 4.16. The right most C labels are quite like those on the 2-gon in Fig. 4.15. In Fig. 4.17, by gluing the designated edges together, we obtain a torus glued to a sphere via a "bridge" or

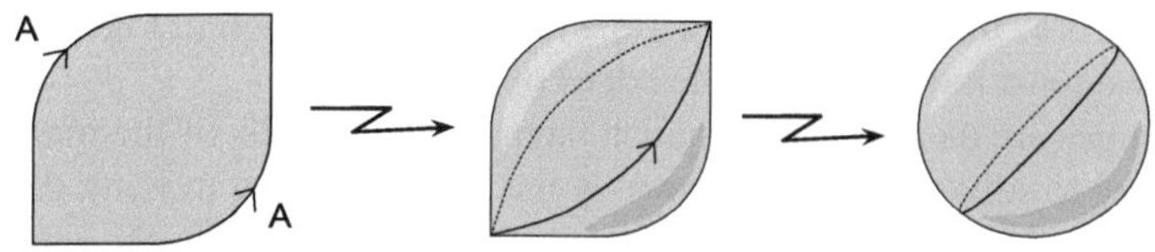

Fig. 4.15 Constructing a sphere from a 2-gon

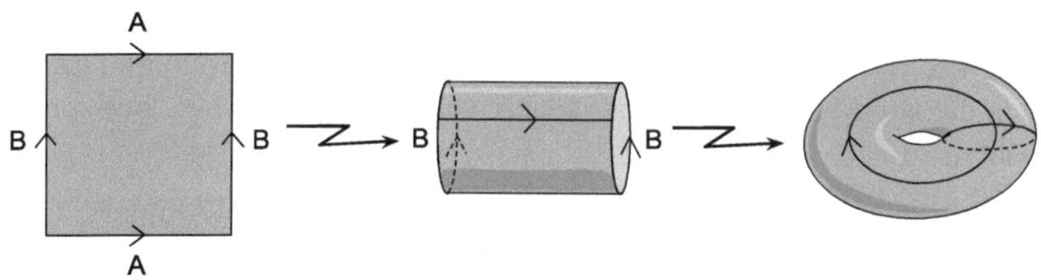

Fig. 4.16 Constructing a torus from a 4-gon

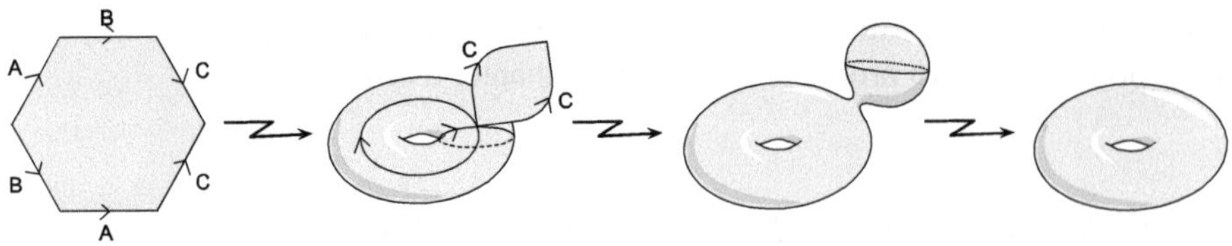

Fig. 4.17 Constructing a torus from a 6-gon

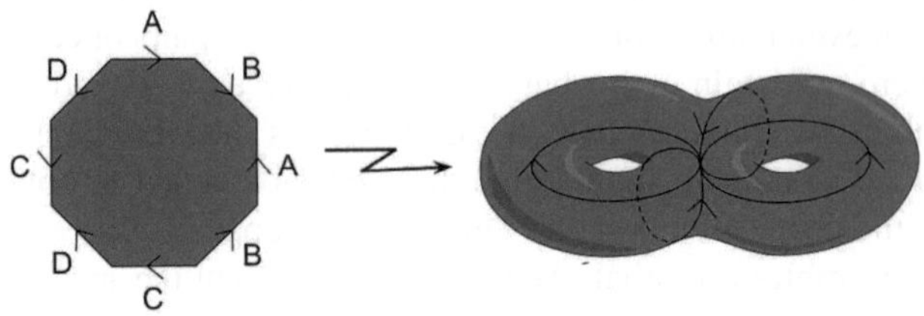

Fig. 4.18 Constructing a two-holed torus from an 8-gon

"neck." As above, we treat this space like it is made from Play-Doh and smash the sphere inward to obtain the torus shown on the far right.

To obtain a torus with two holes, we can glue together the labeled octogon (8-gon) given in Fig. 4.18.

One can describe any connected compact orientable surface by taking a polygon and "gluing together" the polygon's edges in a specified way. When we glue together, or identify, edges, we should do so by making sure that the arrows point in the same direction. Given a polygon P with n-sides, an ***identification space*** is the new space produced by identifying certain edges of P.

While we cannot easily play games on a torus, because we cannot see all the sides at once, we can play games on the corresponding polygon, remembering that certain edges are "the same." Consider the identification space in Fig. 4.19. To solve the maze, we must move through the edges of the square.

These edges are dashed to distinguish them from the walls of the maze. We have started trying to solve the maze, shown by the dotted line, by moving through edge B on the right. This corresponds to coming through edge B on the left. However, this move was not fruitful, because we have reached a dead end.

Exercise 4.4 Going back to start, how would you complete the maze in Fig. 4.19?

Exercise 4.5 In Fig. 4.20a, use the identified sides of the square to find each of the words in the word bank. Words may appear vertically up or down, horizontally, or diagonally.

Word Bank Circle, Line, Cone, Sphere, Torus, Plane, Ball, Geometry, Cylinder, Hyperbolic, Tiling, Distance

Exercise 4.6 Complete the maze in Fig. 4.20 by using the identifications of the sides of the 8-gon.

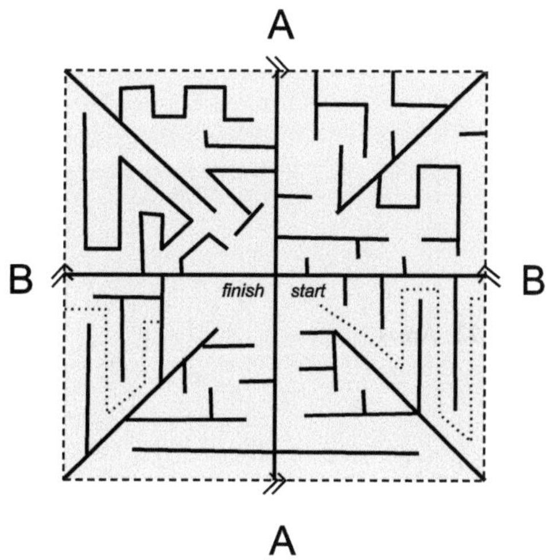

Fig. 4.19 A maze on a torus

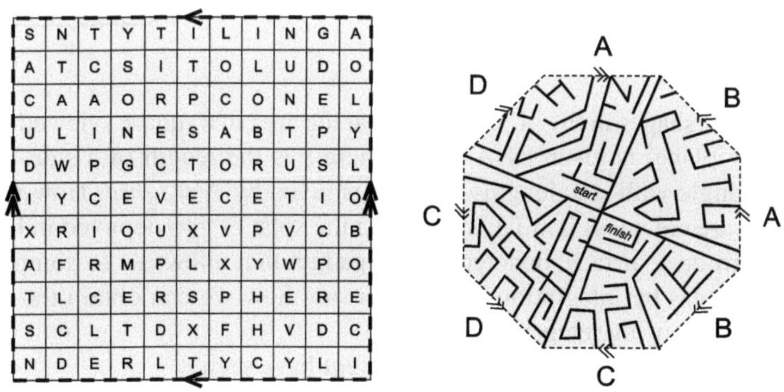

(a) (b)

Fig. 4.20 More games on surfaces. (a) Word search. (b) A maze on a two-holed torus

4.4 Discretized Differential Geometry

A geometer considers the same spaces as a topologist, but with a more refined eye. In particular, a geometer would say that the coffee cup and doughnut in Example 4.6 are different because the cup has more surface area. The ***geometry*** of a space is the set of properties of it which can be measured: distance, area, volume, angles, curvature, etc. The geometric quantity we will focus on is surface area.

Example 4.9 To a topologist, the golf ball and baseball in Fig. 4.21 are the same. To a geometer, on the other hand, these are distinct spaces, because one has greater surface area than the other. □

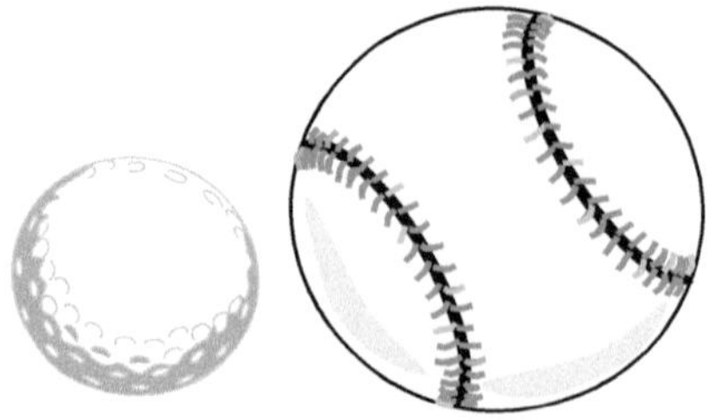

Fig. 4.21 A golf ball & baseball

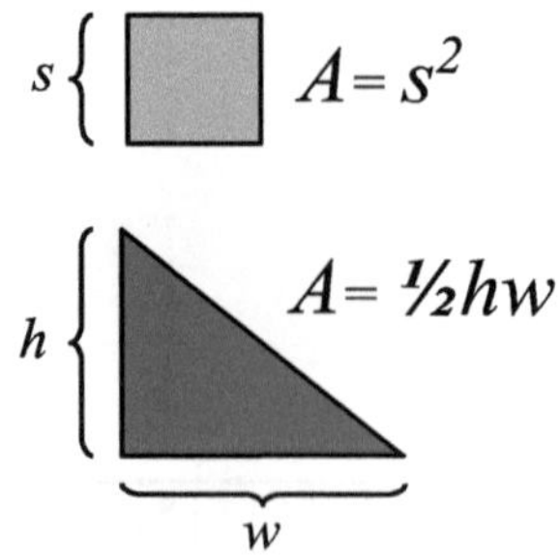

Fig. 4.22 Areas

Consider the shapes in Fig. 4.22. Recall that the area of a square (viewed as sitting inside the xy-plane) is given by $A = s^2$ where s is the length of a side. The area of a right triangle with side lengths h and w is given by $A = \frac{1}{2}hw$. How would we compute the area of a less familiar shape, like the splat of goo depicted in Fig. 4.23? An **area form** on a surface, denoted ω, is a gadget that allows us to compute these less familiar areas by integration. First, let us see how we can obtain the area of a triangle or square using integration.

In a multivariable calculus course, you might have seen

$$Area(\square) = \int_{\square} 1\, dx\, dy \quad \text{and} \quad Area(\triangle) = \int_{\triangle} 1\, dx\, dy.$$

Here $dx\, dy$ is what is the area form and $\int_{\square}$ means integrate over the square. See an illustration in Fig. 4.22.

For instance, if $\square$ is the square with vertices (0, 0), (0, 1), (1, 0), and (1, 1), then

$$\int_{\square} 1\, dx\, dy = \int_0^1 \int_0^1 dx\, dy.$$

Example 4.10 Consider the sphere with spherical coordinates $0 \le \theta \le \pi$ and $0 \le \phi < 2\pi$. Then $\omega = d\theta\, d\phi$ is an area form. We can determine the area of region R in Fig. 4.23 by using the integral

$$\int_R \omega = \int_R d\theta\, d\phi = Area(R).$$

$\square$

Fig. 4.23 Region *R*

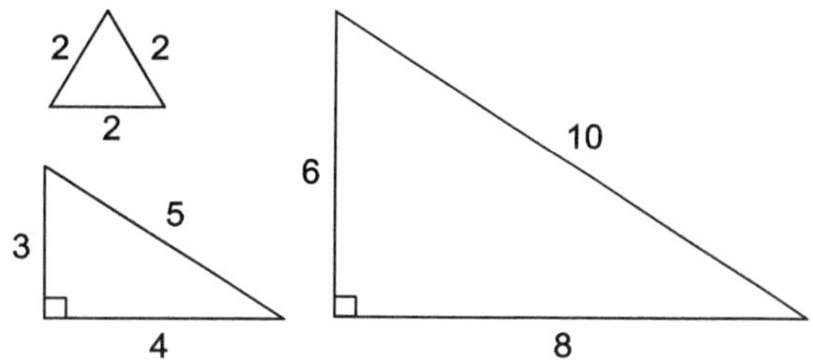

Fig. 4.24 Triangles

Looking at a region R, you may wonder how do you actually determine parameters for it and compute the integral. Indeed, the regions can be gnarly and actually computing the area can be a task in its own right. We will be satisfied with knowing that it can be done rather than extracting an exact numerical answer.

A key problem in geometry is to understand when two surfaces with area forms are geometrically the same. This is a more refined classification question than the one we discussed in the previous section. To illustrate how our notion of "the same" has changed, consider the triangles in Fig. 4.24. All three are different, because they have different side lengths. However, if we only care about the angles of the triangles, then the 3-4-5 and 6-8-10 triangles are "the same." These are called similar triangles: We obtain the larger by scaling the smaller by 2. However, if we only care about the number of sides a shape has, then the three triangles are "the same."

We consider two spaces to be **geometrically the same** if they are topologically the same *and* when we deform one space into the other, we preserve the area of any region R. Miraculously, the following theorem says that for certain spaces, you do not have to check that the area of any arbitrary region is preserved, which certainly is an impossible task since there are infinitely many possible regions on a surface! Instead, we only have to check that the total area of the whole surface is the same.

Theorem 4.2 *Two connected compact orientable surfaces with area forms* (S_1, ω_1) *and* (S_2, ω_2) *are the same if and only if* S_1 *and* S_2 *are topologically the same (meaning they have the same number of holes) and*

$$Area(S_1) = \int_{S_1} \omega_1 = \int_{S_2} \omega_2 = Area(S_2).$$

This theorem is due to the work of Jürgen Moser [12] from 1965. This theorem tells us that we can "concentrate" the area in different ways on the space without changing the geometry as long as the total area of the surface is unchanged!

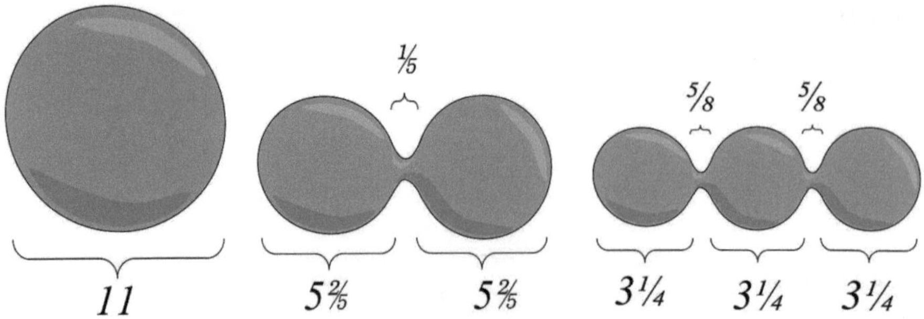

Fig. 4.25 Three geometrically equivalent area forms on a topological sphere

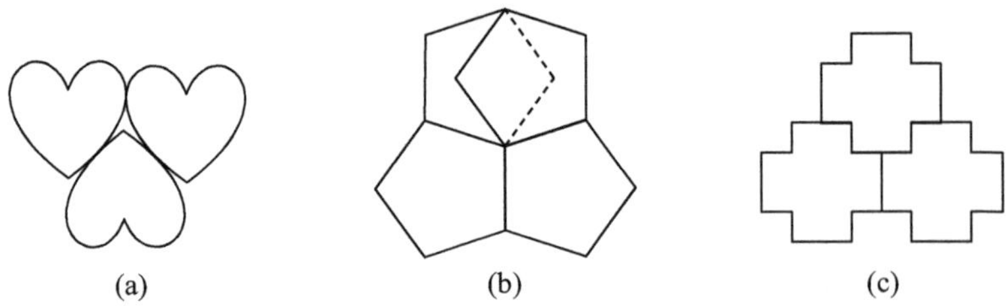

Fig. 4.26 Exploring what is and is not a tiling. (**a**) Hearts. (**b**) Pentagons. (**c**) Crosses

Example 4.11 We can "uniformly" distribute area as shown on the sphere with area 11 in the left of Fig. 4.25. In the center most sphere, we've uniformly concentrated $5\frac{2}{5}$ of the area in a bulb to the left, we have placed a small area of $\frac{1}{5}$ along a connecting bridge, and the remaining area of $5\frac{2}{5}$'s is uniformly concentrated on a right bulb. Since $5\frac{2}{5} + 5\frac{2}{5} + \frac{1}{5} = 11$, this space is geometrically the same as the leftmost sphere. Similarly, the rightmost sphere is also geometrically equivalent to the leftmost sphere. □

4.4.1 *Discretized Area*

We will use tilings to discretize the area of a surface. A ***tiling*** of a surface is a way to cover it using one or more geometric shapes, called tiles, with no overlaps and no gaps.

Example 4.12 The hearts in Fig. 4.26a do not form a tiling of the plane, because they do not fit snugly together. The pentagons in Fig. 4.26b also do not form a tiling because they overlap. The crosses in Fig. 4.26c do tile the plane. □

Given a surface with a tiling, we compute the tile area of a region R by counting the number of tiles that the region touches:

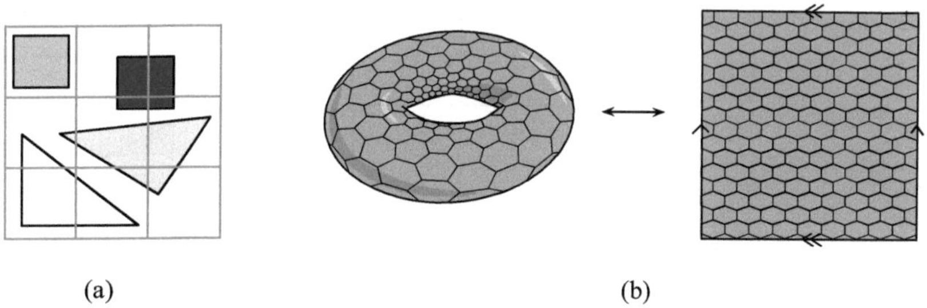

Fig. 4.27 Tile area illustrations. (**a**) Areas. (**b**) The depicted tiling gives a tile-area, or discretized area form, on a torus

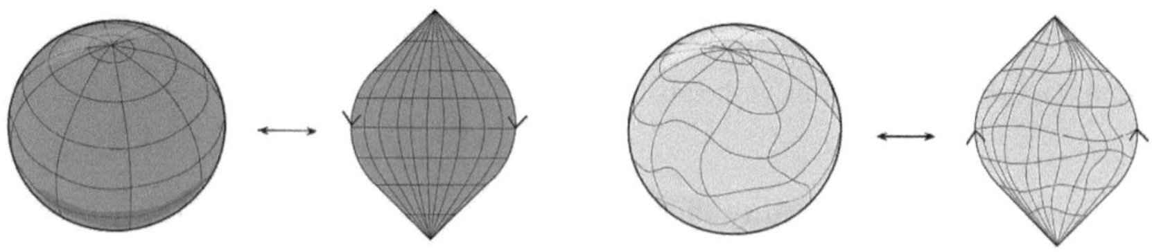

Fig. 4.28 Discretized area forms on the sphere

$$TileArea(R) = \#\{\text{tiles } \tau \text{ such that } \tau \cap R \neq \emptyset\}.$$

Let's return to the example of the area of right triangles and squares in the xy-plane. We can tile the plane with squares. In Fig. 4.27a, the leftmost shaded square touches a single square tile. Thus the tile area is 1. The other shaded square intersects 4 square tiles and so has tile area 4. The tile area of the triangles is 3 and 5, respectively.

Example 4.13 Consider the torus in Fig. 4.27b. This tiling is constructed using 144 hexagons. Thus, the total discretized area (tile area) of this torus is 144. □

Example 4.14 Consider the left sphere in Fig. 4.28. The tiling is constructed using seven latitudinal and ten longitudinal lines. The total discretized area of this sphere is 80.

□

Now that we have a discretized version of an area form; we need to decide what it means for two to be the same. We could say that two discretized areas on a compact oriented surface are the same if they preserve the area of any region R. Would a result like Theorem 4.2 be true for discretized areas? With this proto definition of "the same," if we only check that two tilings have the same number of tiles, i.e., the total discretized areas are the same, we could have a connected region in one tiling become disconnected in another tiling.

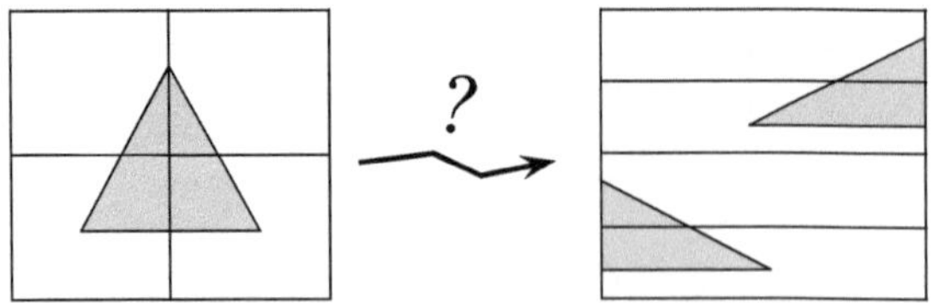

Fig. 4.29 Equivalent area forms?

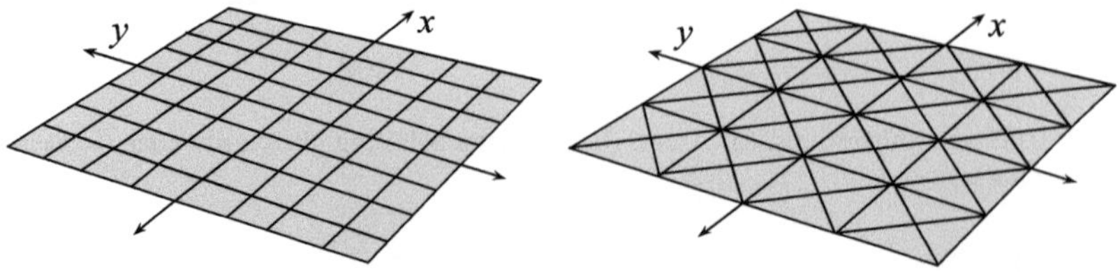

Fig. 4.30 Discretized area forms of the plane

To clarify, consider the tiling of the left square in Fig. 4.29. In order to map the triangle to the tiling on the right, we would have to chop the triangle into two pieces, because there are no places where four tiles meet in the right tiling.

This demonstrates the role topology plays: Two spaces being topologically the same were necessary before they could be considered geometrically the same. Similarly, we need to establish when two tilings are topologically the same. Because connectedness is a desirable property that we want to maintain, we will say two discretized areas are the same if we can map one set of tiles to the other while maintaining adjacencies: Assume tile A_1 maps to B_1 and tile A_2 maps to B_2. Tile A_1 is adjacent to tile A_2 if and only if B_1 is adjacent to B_2. The right sphere in Fig. 4.28 has the same number of tiles as the sphere on the right and the same adjacencies, and thus we consider the corresponding discretized areas to be the same.

Accordingly, the discretized version of an area form is a more rigid object than the original continuous object.

Exercise 4.7 Consider the xy-plane with the standard (familiar) area. The tiling by squares and the tiling by right triangles – shown in Fig. 4.30 – are possible discretized areas. Compare the adjacencies of each tiling to see that they give different tile areas.

4.4.2 Surfaces with Infinite Ends

In my research (see [8] or [9], for instance) I work on generalizing Theorem 4.2 to a type of non-compact surface called a surface with an end. The ends that I work with are places on an otherwise compact surface that extends out infinitely far and resemble a familiar space. For example, the tori in Fig. 4.31 each have an end. The

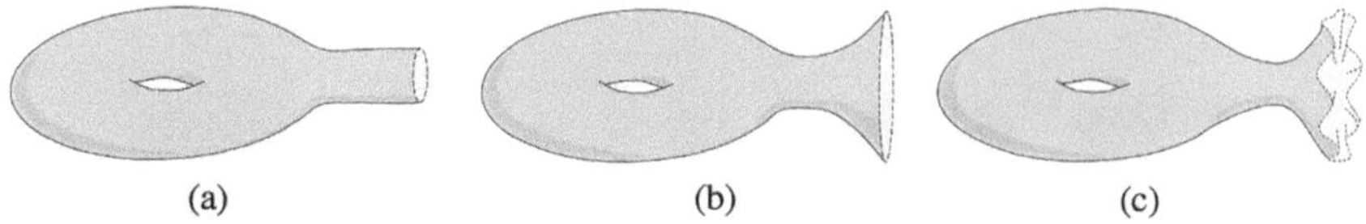

Fig. 4.31 Tori with ends. (a) Cylindrical. (b) Conic. (c) Hyperbolic

first one's end is a cylinder, the second's is the wide end of a cone, and the final one's is hyperbolic space.

My first step is to take two surfaces with the same type of end and glue the two ends together, thus creating a compact oriented surface. In particular, we begin with the two surfaces with ends and add a circle to the edge of each end – this is called compactifying – to make a surface with boundary. We then glue the two circles together.

To distinguish the two sides, we consider one to have positive area and the other to have negative area. This is comparable to how integrating a real function computes area. Remember that the area of a function whose graph lies below the x-axis is negative. Since an end extends infinitely far out into space, surfaces with an end have infinite area. By taking one side to have negative area and one side to have positive area, we can obtain a finite total area for this new space.

In this construction, it is important to only glue cylindrical ends to cylindrical ends, conical ends to conical ends, and hyperbolic ends to hyperbolic ends, because these spaces extend infinitely far in different ways. To illustrate, consider the growth of the functions $f(x) = x$, $g(x) = x^2$, and $h(x) = x^3$. Notice that

$$\lim_{x\to\infty} f(x) = \lim_{x\to\infty} h(x) = \lim_{x\to\infty} g(x) = \infty.$$

If we add $f(x) + 4$ to $-f(x)$, then

$$\lim_{x\to\infty} (f(x) + 4) + (-f(x)) = 4.$$

As long as we pair functions with the same asymptotic growth, we can add them together with one taken as negative and obtain a finite number in the limit, i.e., like we did with $(f(x) + 4) + (-f(x))$. However, if we mismatch behavior, like in the limit

$$\lim_{x\to\infty} f(x) + (-h(x)) = -\infty,$$

we still have something infinite.

Example 4.15 Consider the left two spheres in the upper row of Fig. 4.32. The light grey sphere has area $\frac{1}{2}$. The dark gray sphere has area $-\frac{1}{2}$.

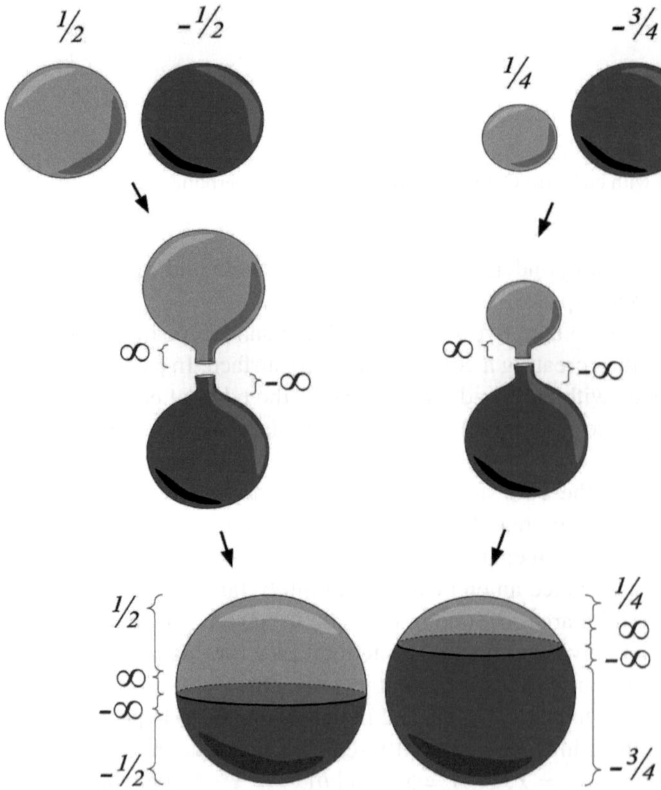

Fig. 4.32 Two generalized area forms on the sphere

In the next row, we see a hole poked in both spheres and stretching the sphere to make a cylindrical end. The light gray sphere's end has area ∞. The dark gray sphere's end has area $-\infty$.

In the final row, we have our final space, obtained by gluing the two spheres with ends together. The top half, shown in light gray, will have positive area. The bottom half, shown in dark gray, will have negative area. The top hemisphere, away from the equator, has surface area $\frac{1}{2}$. As described above, the region in the top hemisphere near where we glued the two pieces together has area ∞. The bottom hemisphere, away from the equator, has surface area $-\frac{1}{2}$. The region in the bottom hemisphere near the circle where the spaces are glued together has negative infinite area. Importantly, we glued two spaces together that had the same type of end – be it cylindrical, conic, or hyperbolic. In this special case, the ∞ and $-\infty$ have the same magnitude and cancel to 0. Further, $1/2 + (-1/2) = 0$ and thus the total area of the left sphere is 0.

Next, consider the right two spheres in the upper row of Fig. 4.32. The light gray sphere has area $\frac{1}{4}$. The dark gray sphere has area $-\frac{3}{4}$. In the next row, we see a hole poked in both spheres and stretching the sphere to make a cylindrical end. The light gray sphere's end has area ∞. The dark gray sphere's end has area $-\infty$. In the final row, we have our final space, obtained by gluing the two spheres with ends together. The upper hemisphere away from the circle has area $\frac{1}{4}$, and the region in the upper hemisphere near the circle has area ∞. The lower hemisphere has area $-\infty$ near the circle and $-\frac{3}{4}$ away from the gluing. Thus, the total area of the right sphere is $\frac{1}{2}$. □

Consider the sphere with spherical coordinates $0 \leq \theta \leq \pi$ *and* $0 \leq \phi < 2\pi$. *If the ends we originally considered were cylindrical, then*

$$\omega = \frac{1}{\sin(\theta - \pi/2)}\, d\theta\, d\phi \qquad \text{and} \qquad \omega = \frac{1}{\sin(\theta - \pi/3)}\, d\theta\, d\phi$$

denote the generalized area form on the left and right spheres, respectively. Note that we have $\sin(\theta - \pi/2) = 0$ *for* $\theta = \pi/2$, *which defines the circle where we glued the two pieces to make the left sphere. Similarly,* $\sin(\theta - \pi/3) = 0$ *for* $\theta = \pi/3$, *which defines the circle where we glued the two pieces to make the right sphere.*[5]

These spaces and areas are novel, so there is no corresponding discretization theory. We are currently expanding the continuous theory. Developing a corresponding discretization is an open and available problem.

4.4.3 Open Questions for You to Ponder

We leave you with an open question: What would be an appropriate discretization for these generalized area forms?

To Learn More To develop more intuition about topology, I highly recommend Jeffrey Weeks' book The Shape of Space [15]. There are several relatively new text books discussing discretized differential geometry, for instance, Alexander Bobenko's book Discrete Differential Geometry [2].

References

1. Birman, J.S., Eisner, J.: Seifert and Threlfall, A Textbook of Topology, vol. 89. Academic Press, New York (1980)
2. Bobenko, A.: Discrete Differential Geometry. Springer, Berlin (2008)

[5] Yes, we just divided by zero! Sometimes it makes perfect and total sense to do so; you just have to be careful.

3. Bolzano, B.: Rein analytischer Beweis des Lehrsatzes, das zwischen je zwey Werthen, die ein entgegengesetzes Resultat gewähren, wenigstens eine reelle Wurzel der Gleichung liege. gedruckt bei Gottlieb Haase (1817)
4. Cauchy, A.L.B.: Cours d'analyse de l'École Royale Polytechnique. Debure (1821)
5. Crane, K., Wardetzky, M.: A Glimpse Into Discrete Differential Geometry. Notices of the AMS, vol. 64(10) (2017)
6. Lakoff, G., Johnson, M.: Metaphors We Live By. University of Chicago Press, Chicago and London (1980)
7. Lakoff, G., Núñez, R.: Where Mathematics Comes from, vol. 6. Basic Books, New York (2000)
8. Lanius, M.: A classification of star log symplectic structures on a compact oriented surface. J. Symplectic Geom. **17**(6), 793–1833 (2019)
9. Lanius, M.: Symplectic, Poisson, and contact geometry on scattering manifolds. Pac. J. Math. **310**(1), 213–256 (2021)
10. Lanius, M.: On Conceptual Metaphor in Cryptology Education. Problems, Resources, and Issues in Mathematics Undergraduate Studies, p. 12 (2023)
11. Lanius, M.: Through the looking glass, and what algebra found there: historically informed conceptual metaphors of algebraic substitution & gaussian elimination. Br. J. Hist. Math. **32**(2), 141–157 (2023)
12. Moser, J.: On the volume elements on a manifold. Trans. Am. Math. Soc. **120**(2), 286–294 (1965)
13. Sperner, E.: New proof for the invariance of the dimension number and the area. In: Essays from the Mathematical Seminar of the University of Hamburg, vol. 6, pp. 265–272. Springer, Berlin (1928)
14. Walsh, T.: Timeless Toys: Classic Toys and the Playmakers Who Created Them. Andrews McMeel Publishing (2005)
15. Weeks, J.R.: The Shape of Space. CRC Press, Boca Raton (2001)

Chapter 5
The Scholarship of Teaching and Learning

Rebecca Swanson

Abstract In this article, we will introduce the reader to a type of research called the Scholarship of Teaching and Learning (SoTL), in which teachers study, analyze, and publicly share information about the teaching and learning of their own students. Examples of SoTL studies will be presented, including a summary of the author's work in Linear Algebra and Calculus courses. Although SoTL differs from traditional mathematical research, content knowledge of mathematics is necessary for the planning of content delivery, for recognizing problems in student learning of that content, and for conjecturing solutions to those problems. In this article, a knowledge of standard introductory mathematics courses is assumed, particularly those of linear algebra, calculus, and some introductory statistics.

Keywords Scholarship of teaching and learning · SoTL · Flipped learning · Educational research · Linear Algebra · Calculus · Evidence-based teaching · Educational research · Coordination · Alternative assessment

5.1 Biography/Career Trajectory

As a high school student in rural North Dakota, I enjoyed and experienced success in all of my classes. However, it wasn't until a friend recognized that many of my study hall hours were spent assisting classmates with their mathematics homework that I became aware of how much I enjoyed helping others make sense of math. That same friend told me of a summer mathematics and science program at North Dakota State University called Governor's School. I applied to the program, was accepted, and had my first experience working with like-minded people to learn and understand challenging and abstract mathematical ideas. From then on, I was hooked and knew that I wanted to continue to study mathematics in college. I enrolled at Dakota

R. Swanson (✉)
Colorado School of Mines, Golden, CO, USA
e-mail: swanson@mines.edu

A. Brisbin et al. (eds.), *Research Connections*, Association for Women in Mathematics Series 36, https://doi.org/10.1007/978-3-031-81217-0_5

Wesleyan University (DWU), a small liberal arts college of about 650 students in nearby Mitchell, South Dakota. I was unaware at the time that a math department consisting of three faculty members was considered very small. Although I may not have had the course options of my peers at larger institutions, the support, guidance, and opportunities I received from my professors were invaluable.

My advisor, Mike Catalano (who will always be Dr. C. to me!), told me I should apply to a program at Carleton College called the Summer Math Program for Women (SMP). I had never heard of it but thought it sounded like fun. I was accepted and headed off to Northfield, Minnesota, with no idea what to expect. I remember the excitement of being part of a group of clever and interesting women who enjoyed mathematics. Erica Flapan and Karen Brucks served as the faculty instructors that summer. Their classes were new and challenging, and I remember, especially in Karen's class, feeling overwhelmed. I would attend her office hours, and she would patiently (and repeatedly) guide me through my misunderstandings. She also encouraged me to ask questions in class. I had been scared to do so, both fearing that I was the only one who didn't understand and not knowing what to ask. Karen helped me see that I was not alone and gave me script for asking questions when I felt lost. She told me to raise my hand and to simply ask "Can you go over that again?". I have shared her lesson with my own students by reminding them that they are not alone and providing them with a similar "script." Given my struggles, I had some doubt that summer about whether I was smart enough to continue to study math. It was also Karen who helped alleviate that fear. On an SMP weekend outing while tubing down a river, Karen asked me if I was planning to go to graduate school. I remember that moment vividly. No one had ever asked me that question, and in asking that question, she conveyed that **I** could go to graduate school. It was then that the idea of pursuing a mathematics graduate degree felt like a real possibility to me.

During SMP, I was introduced to Paul Humke of St. Olaf University and director of the Budapest Semesters in Mathematics (BSM) program. Paul's presentation sold me on the idea of spending a semester in Budapest, Hungary, learning mathematics, and I was accepted for the following spring. Again, the experience was both challenging and incredible. I loved learning and spending time with my classmates exploring a new place and culture. I returned home at the end of BSM invigorated but still unsure of what role I wanted mathematics to have in my life. In fact, I spent the next summer doing something entirely different: I helped run a homeless shelter and free lunch program for low-income children. Although this experience wasn't mathematical, it was important for me to share my time with others. In this role, I often worked a late night/middle of the night shift, where I found I was entertaining myself by doing math problems! I left that summer with a new perspective and the realization that I wanted to continue with mathematics. What I did not have was any sense on what I should do upon graduation. Although I enjoyed helping others learn, being a high school teacher wasn't appealing to me. I knew I enjoyed mathematics but hadn't done any math research. I had heard of undergraduate research programs (REUs), but if I was going to graduate in the spring, I wouldn't be eligible for most of them. Therefore, I made the decision to spend an extra year in college. I spent my “first” senior year working on an independent research project with Dr. C., and I attended the Indiana University REU, the following summer. These experiences demonstrated that I enjoyed mathematics research in addition to helping others learn mathematics. It was in reflecting upon Dr. C’s job that made me finally realize that I wanted to be a math professor. He seemed to have a great time teaching a variety of classes and working with students. I felt that I needed to learn some more mathematics to prepare for graduate school, so I spent the fall of my fifth year of college at the Mathematics Advanced Study Semesters program at the Pennsylvania State University and then returned to BSM for the spring. That final year, I applied to graduate programs and decided to attend Indiana University, due in large part to my positive REU experience and a graduate fellowship offer.

At Indiana University (IU), I studied Combinatorics under Patricia Hersh. As an undergraduate, my favorite subject was combinatorics. I arrived at IU open to the idea of studying something else, but combinatorics remained the most interesting. Graduate school was certainly challenging – working on mathematics research is very different from doing homework! Some of my favorite parts of grad school were my first experiences teaching. However, I also struggled with the traditional academic notion of “success.” Although my goal upon completing graduate school was to be a professor, ideally in a small liberal arts college, I felt that there was a professional expectation to attain a faculty position that prioritized research over teaching. In particular, it seemed that to be successful, a graduate student would finish graduate studies, complete postdoctoral research, and then attain a faculty position in which mathematics research and publications were a large component of the job. That path wasn’t appealing to me, and I struggled with my mathematical identity, often feeling like an impostor. It took time to gain confidence in my professional goals, but by the time I applied for jobs, I targeted positions that clearly valued teaching but still had some research expectations.

On a more personal note, it was at IU that I met my husband, Stephen Pankavich. Because we were both mathematicians, we had what is referred to as a “two body problem,” meaning that we needed to find a location in which we could both find jobs. As an additional challenge, we had different career goals, as Steve wanted a more traditional research/teaching academic position. After a couple of years in different states, we were successful in finding positions at the Colorado School of Mines (Mines). What made Mines ideal for us was that the university has both a traditional job track and a teaching-oriented track. I am now a Teaching Professor, and my primary focus is teaching. My research on teaching (the subject of this article) is considered valued scholarship for my position. Among other nonteaching tasks, I support students outside of the classroom as co-founder and co-advisor to our local Association for Women in Mathematics chapter, have advised our university Putnam Team, and advise a cohort of women STEM scholars at Mines. We are happy to be in Colorado and welcomed our daughters, Eloise and Diana, in 2017 and 2019, respectively.

Over the course of my training in mathematics and my academic career, I have learned some lessons along the way. First, feeling like an impostor is not unusual and is often experienced by those who are underrepresented in their profession.

I have felt it at times throughout my training and career, and while it is neither enjoyable nor comfortable, there are ways to combat impostor feelings. One method, which I credit to my former SMP teacher, Erica Flapan, is to recognize that there are many different ways to be successful in mathematics. More generally, it is important that you know and determine what it is that defines your success. In my case, professional success meant having an academic position that valued both teaching and scholarship related to teaching. I have also found that "evidence" builds confidence and combats impostor feelings. For instance, every time I complete something that I used to find intimidating, I have additional evidence of success (or at least survival). The other related, big lesson I want to share is the value of saying yes to new experiences, both personally and professionally. You never know what you might learn or which relationships you may develop from doing something new. As you will read about in this article, jumping into an interesting project with two colleagues led to a new scholarship experience; local, regional, and national presentations; and publication of our work. Of course, applying to and attending SMP brought me great friendships and planted the seed that led me to graduate school and my current career, and I am forever thankful that I am part of the SMP family.

5.2 Introduction

As an undergraduate mathematics student, one learns of the numerous career opportunities afforded to mathematicians. From actuaries, to consultants, to national security analysts, to statisticians, the options are seemingly endless. Both the Mathematical Association of America's book, *101 Careers in Mathematics* [41], and the site "We Use Math" [45] provide many such examples. However, in my experience, the career option with which students of mathematics are most familiar is that of teaching mathematics. It is the one to which we are exposed throughout our pre-collegiate and collegiate experience as we interact with our own teachers of mathematics. Most postsecondary academic positions require some mix of responsibilities that include both teaching and research. In this article, we will learn about the Scholarship of Teaching and Learning (SoTL), in which faculty members systematically study their own teaching and student learning and publicly share their results. Some mathematics faculty who work in SoTL do so in addition to more traditional mathematical research, while the research of others may lie exclusively within SoTL. Our goal here is to introduce the reader to SoTL, the necessary background, and how such research is performed. To this end, we present illuminating examples of SoTL questions and studies, including the author's own forays into SoTL. Some of the examples presented herein are simplified with the goal of improving understanding of nonexperts.

5.2.1 Scholarly Teaching

Although teaching is often the aspect of a faculty position that is most visible to students, the training of mathematicians to teach mathematics is often quite minimal. Having already earned a graduate degree, we are experts in our own research areas, and the traditional assumption is that this expertise translates into an ability for us to transfer our expert knowledge to students through teaching. Unfortunately, as you may have experienced as an undergraduate student, this is not always the case. Mathematics faculty typically have their first experience teaching in graduate school, and in many cases, they do so with minimal training [35]. At times, such minimal training can last an afternoon, a couple of days, or possibly a week. Fortunately, an increasing number of graduate schools are offering, and, in many cases, requiring, teacher training, as evidenced both in [35] and the Mathematical Association of America's "College Mathematics Instructor Development Source" (CoMInDS) program. In particular, the National Science Foundation has provided CoMInDS with funding to train mathematics faculty to develop or enhance teacher training programs for their own graduate students, as those graduate students will often become future faculty members [11]. In addition to the improvements in training for graduate students who will become mathematics faculty, many universities have developed centers for teaching and learning that support faculty interested in becoming better teachers.

So, what does it mean to become a "better teacher?" If we agree that the goal of teaching is student learning, then better teaching should translate into a measurable improvement in student learning and our understanding of student learning. There is a large body of work on effective teaching and what it means to be an effective teacher. For instance, in [16], Freeman, et al. demonstrated the importance of student engagement in learning. They analyzed 225 different studies on teaching and learning to measure the impact of active learning versus the more traditional teaching method of lecturing. Although the definitions of "active learning" in these studies varied, they contained a common thread of increased student engagement. They showed that on average, student performance in an active learning classroom increased by approximately a half of a standard deviation under active learning and students in lecture courses were almost twice as likely to fail as students in active learning classes. This is one example of what the literature tells us about teaching. Numerous other examples exist, and as teachers of mathematics begin to learn about and implement effective teaching strategies and reflect upon their experiences, they are becoming "scholarly teachers." In [36], Potter and Kustra provide the following definition of scholarly teaching:

> **Scholarly teaching** is defined as teaching grounded in critical reflection using systematically and strategically gathered evidence, related and explained by well-reasoned theory and philosophical understanding, with the goal of maximizing learning through effective teaching.

In short, scholarly teachers learn about and implement effective, evidence-based instruction methods in their classes and then reflect upon their experiences with the goal of improving student learning.

Before introducing SoTL, it is necessary to first examine scholarly teaching, because although we will see that SoTL and scholarly teaching are distinct from one another [36], for many of us who have worked in SoTL, we first became scholarly teachers. For many of us who have completed SoTL projects, myself included, our paths began by first reflecting upon our own teaching and questioning its effectiveness. As an example, I will share a bit of my own journey. I was one of the many graduate students who received minimal training (an afternoon seminar) before entering the classroom. I believed that I had a talent for explaining mathematics, and my first experiences in the classroom – and my first student evaluations – seemed generally positive. However, the more I taught, the more I began to question what it really meant to be a "good teacher" and how I knew whether and what my students were learning. I began to attend teaching-related workshops at conferences, audited a science education class, and learned more about effective teaching. Upon graduation, the collegiate mathematics professional development group Project NExT, which stands for "New Experiences in Teaching" [38], provided me additional experiences in learning about best practices in teaching. After becoming a mathematics faculty member, I increasingly began to implement the ideas that I learned about, gather my own evidence on the success of my endeavors, and reflect upon that evidence with the goal of making continued improvements to my classes. By engaging in these activities, I became a scholarly teacher.

5.2.2 *What Is SoTL?*

Jackie Dewar and Curtis Bennett of Loyola Marymount are the authors/editors of "Doing the Scholarship of Teaching and Learning in Mathematics," [14], a book aimed at helping college faculty members learn about and develop their own SoTL projects. Dewar and Bennett have had a tremendous impact on growing the SoTL community in collegiate mathematics and increasing the visibility of SoTL in that community. In addition to their text, which has had a great influence upon my own work, they have presented talks, contributed to panels, and developed workshops on SoTL. In their book [14], Dewar and Bennett describe a path similar to my own in which one moves from Scholarly Teaching to the Scholarship of Teaching and Learning. They define the Scholarship of Teaching and Learning as follows:

> **The Scholarship of Teaching and Learning (SoTL)** is the intellectual work that faculty members do when they use their disciplinary knowledge (in our case, mathematics) to investigate a question about their students' learning (and their teaching), submit their findings to peer review, and make them public for others to build upon.

When teaching mathematics, we regularly encounter teaching and learning problems: *Is one teaching method more effective than another? How do I get my students to prepare for class? What are my students doing to prepare for class? How do I help my students become better mathematical writers? Is online homework as effective as written homework in learning mathematics?* The list of such problems can quickly become quite long! A scholarly teacher encountering such problems would begin by consulting previous studies to see if others have encountered the problem and, hopefully, found a suitable solution. If there seems to be a body of evidence that indicates an effective solution, the scholarly teacher would attempt to implement it, gather evidence, and reflect upon the process. However, in many cases, there might be no solution or only partial solutions in the literature. A solution can be partial for a variety of reasons. A large number of variables could have effects on the collected data and the interpretations of that data. For instance, if someone at an engineering university found that online homework was as effective as written homework in learning mathematics, a faculty member at a small liberal arts college may end up with different results. Or if you were asking about what it is that students do to prepare for class, a study of a general education math course might not apply to your upper-level major course. Additionally, some of the research on teaching and learning that we consult originates outside of mathematics, and we may want to know similar results would arise within the mathematics setting. Of course, it is also possible that no one has considered the problem that we face. In either case, if we design a study, implement the design, collect evidence, and make our findings public (and peer-reviewed!), we have moved from scholarly teaching to SoTL.

To get an idea of the types of questions or problems that faculty might consider as SoTL projects, let us look to a somewhat recent Joint Mathematics Meetings (JMM). For roughly the past decade, JMM has been host to contributed paper sessions on SoTL. Here are some examples of SoTL projects presented in the 2018 "Session on the Scholarship of Teaching and Learning in Collegiate Mathematics," organized into some observed themes:

- Student thinking
  A common theme of some SoTL projects is related to developing a better understanding of student thinking. For instance, Girija Sarada Nair-Hart of the University of Cincinnati was interested in how non-math majors developed their mental models of various calculus concepts [31]. As another example, Tharanga M. K. Wijetunge of Lyon College and Kirthi Premadasa of University of Wisconsin-Baraboo/Sauk County wanted to know how students select the order in which they complete questions on mathematics exams [47].
- Mindset
  In recent years, there has been a body of research related to fixed and growth mindsets and their relationship to learning [4]. Someone with a fixed mindset believes that his or her intelligence is "fixed." In the case of mathematics, a person with a fixed mindset would believe that an individual can either learn it or is unable to learn it. Someone with a growth mindset in mathematics would believe that it is possible for anyone to learn math, and that intelligence can grow. Having

a fixed mindset can be damaging to learning, so it is therefore not surprising that a number of SoTL projects have risen in relation to this concept. Yevgeniya Rivers of University of New Haven posed the question "How is growth mindset and metacognitive self awareness linked in developmental math students?" as the basis for a SoTL project [39]. Similarly, Ben Hartlage, David Higginbotham, Ian Pierce, and Beth Schaubroeck of the United States Air Force Academy were interested in how grit and diligence as personality traits were related to success in mathematics courses [20]. Finally, Lauren Klamerus, Christina Carlson, and Amanda Harsy of Lewis University were interested in Mastery Based Testing, in which students can only receive full credit for problems once they have demonstrated mastery of the related concept [24]. Their hope was that this would motivate students to learn from their mistakes and counteract a fixed mindset of learning.

- Classroom interventions
  A classroom intervention refers to something you do in order to address a teaching problem that you are having. Another common theme of SoTL projects is related to testing a classroom intervention of some sort to see if it works better than another intervention. For instance, at the University of Oklahoma, a new active learning model was implemented in precalculus and trigonometry courses, and Keri Kornelson of the University of Oklahoma presented results related to the impact of this model on math attitudes of students, grades in current and future courses, and the mindsets of students [25]. Lisa Bromberg and Michael Seminelli of the United States Military Academy–Westpoint were interested in the effect of digital versus paper quizzes on student performance in a first year collegiate mathematics course [6]. Andrew G. Bennett and Ian Manly of Kansas State University were interested in the problem of retaining students in STEM (Science, Technology, Engineering, and Mathematics) disciplines [2]. They presented results related to the introduction of a bailout course that a student could drop to in Calculus I with no penalty or cost if the first exam went poorly, and its effects on retention. As a final example, Alison Lynch of California State University–Monterey Bay wanted to see if students reviewing their peers' papers (in a structured way) improved their performance [28].

5.2.2.1 SoTL Taxonomy

All of these projects can be classified into a few different types. A taxonomy is a scheme for classification, and a variety of taxonomies for SoTL questions have been proposed (e.g., [32], [30]). One of the more user-friendly schemes was proposed by Hutchings in [23] and modified by Dewar and Bennett in [14]. On page 14 of [14], the following taxonomy and descriptions are provided:

- *What works?*
  This question asks whether a particular teaching method, assignment, or approach is effective.

- *What is?*
  This question seeks detailed information about what students are really doing in a particular situation.
- *What could be?*
  This question examines a case where something really interesting happened or a situation involving something the instructor is passionate about or has an opportunity to do. This question type aims to show what is possible, so it is sometimes referred to as a *vision of the possible*. (Note: The *What could be?* question has the goal of approaching novel ideas with the focus on the idea. It can lead to a *What works?* question, in which the focus would be on the effectiveness of the idea.)

This taxonomy can be helpful in formulating a SoTL question and study. In regards to the above examples, those under the "What are students thinking?" bullet most likely fit into the *What is?* category, as does [39]. Many of the remaining fall under the *What works?* category with the exception of [2], in which parts fall in the *What could be?* category. In fact, the lines often blur between categories, and, according to Dewar, the taxonomy should be used as more of a guide when formulating a SoTL question of study as it can help determine the types of evidence one chooses to gather. According to [14] and [23], many faculty begin in the *What works?* category but are led to questions in the *What is?* category. For instance, in [47], the investigators are interested in improving success on mathematics exams, and they suspect the order in which a student selects problems for the exam might influence exam performance. Before designing an intervention to help students, they first need to know what students are already doing on exams in regards to the selection of problems. In the next sections, after discussing the steps of developing and completing a SoTL project, we will present some examples of SoTL work in more detail. First, we address what SoTL is not.

5.2.2.2 What SoTL Is Not

Before reading this article, some readers may have heard of "Mathematics Education Research" (MER), or "Math-Ed," as some universities have undergraduate or graduate programs or concentrations in "Mathematics Education." We should point out here that there are members of the MER community that focus on K-12 learning and others that focus on postsecondary learning. Many of the big goals of postsecondary MER and SoTL are similar. For instance, both SoTL and postsecondary MER researchers are interested in how students think about mathematical ideas and ways to support student learning. It is natural at this point to ask what the difference is between postsecondary MER and SoTL. Through conversations, experience, and reading (e.g., [27]), I have observed the following differences but should note that this list is not complete and that there are a number of subtleties between the two fields that we will not address here. Specifically, postsecondary MER researchers and SoTL researchers have different training, (sometimes) different types of jobs, and often different approaches to solving problems. A SoTL researcher possesses a primary expertise in mathematics and has gained additional expertise in effective teaching and completing a project related to their own teaching problems. A typical SoTL researcher began their career by completing a graduate degree in mathematics, meaning that their graduate research was focused on a mathematical topic, not the teaching or learning of mathematics. A SoTL researcher then became part of the SoTL community by working as a faculty member in a department who also studies teaching problems in a systematic way and publicly shares the results of the studies. A SoTL researcher often does mathematics research in addition to SoTL work. A postsecondary MER researcher has expertise in educational theories of learning and the application of those theories in addition to some mathematical expertise. A postsecondary MER researcher will typically have earned a master's degree in Mathematics, as well as graduate degree in Mathematics Education, meaning that their thesis work was related to mathematics teaching and learning instead of a purely mathematical project. A MER researcher may work in either a math department or an education department. This means that they may teach either mathematics classes or mathematics education classes, or some mix. Because of their expertise in theories of learning, postsecondary MER researchers will be able to apply those theories to design studies, and such knowledge will help the researcher interpret and explain results. In short, the theory that a MER researcher studies not only gives the researcher guidance on which data to gather but will also allow the researcher to interpret the results. To summarize, SoTL and MER are distinct research fields that certainly have some overlap. The purpose of this article is to introduce you to SoTL, but if the reader would like to learn more about mathematics education research, [40] can provide an introduction.

5.2.3 How to Do a SoTL Project

After researching a teaching problem, we work to refine our SoTL question using the SoTL taxonomy as a guide. Once we have a question, we must next formulate a plan for which data we will collect. Once the data has been collected, it must be analyzed. Finally, to truly be a SoTL project, we must subject our results and analysis to peer review by either publishing or publicly sharing the work with peers. This process contains a number of challenges for mathematicians:

- Reviewing literature
  We begin a SoTL project by reading and learning what others have done when encountering the same, or similar, teaching problems. Identifying and obtaining such research are new to many mathematicians. Having trained within our particular disciplines, we are quite familiar with mathematics journals, but we often have less experience with SoTL research outlets. One mathematics journal that regularly publishes SoTL projects is PRIMUS [37]. However, mathematics SoTL projects can also be published in multi- or interdisciplinary journals. A list of great resources can be found in Chapter 2 of [14].

- Planning a project
  Once we have a question (or questions) we want to study, we must form a plan for what we will do and which data to collect. Especially in the case of a *What works?* question, planning our study has some experimental design challenges. It is rarely possible to construct a true control group to which the experimental group may be compared. Even if we have the opportunity to compare results across two (or more) groups of students, there can be additional explanations for differences in results. For instance, we might be unable to teach both groups of students. Or even if we can teach both groups, we may unconsciously bias the results by treating the groups differently. Additionally, if we believe strongly in a teaching intervention, we may find it unethical to not apply that intervention to a "control" group. Another experimental design challenge is that in many cases, the sample sizes might be too small, especially when working at a smaller university or in upper level major courses. There are various attempts we can make to combat these challenges. For instance, we might compare results from previous semesters if we can justify doing so. Or we might aggregate data across semesters until a sample size is sufficiently large.

  Once we have planned a study, in order to make the future results of that study public, we must obtain approval (or "exemption") from an Institutional Review Board (IRB). When studying student learning, we are working with human subjects, another novelty for many mathematicians. Luckily, there is usually institutional support to aid faculty in this process and to provide them with IRB training.
- Data collection and interpretation
  An additional challenge may arise when we formulate a plan for data collection. As mathematicians, we often rely heavily on quantitative data, but when engaged in SoTL, we must accept both the limitations of quantitative data and the benefits of using more qualitative data. There are many examples of quantitative data that we may collect. For instance, we could collect data related to student performance, such as scores on homework and test problems. Alternatively, we could collect student survey data, such as student ratings on aspects of a particular course model, their ratings of their confidence in being able to complete course-related tasks, or their reported behavior, such as time spent on a task. Mathematicians are often more comfortable working with such quantitative data, and that has certainly been true in my case. The caveat here is that we must be careful about inferring too much from quantitative data. Even quantitative data can be very subjective, as there is a choice in which questions we ask, how we analyze responses, and which data we choose to gather [33].

  Mathematicians also have very little experience collecting or working with more qualitative data. Some examples of qualitative data that may be collected in a SoTL project include student responses to questions (e.g., exam responses, homework responses, survey responses) and videos of students discussing or presenting mathematics. Often these are tools of social science or educational research and require some additional training for most mathematicians. Our goal here is not to teach the reader how to collect or work with qualitative data but rather to demonstrate that such data exists and can be quite useful. One advantage of using such data is that it can uncover student thinking in ways that quantitative data cannot.

Finally, many mathematicians have learned some elementary statistics but may feel uncomfortable interpreting or analyzing data. This is a skill that can be learned in working in SoTL.

- Publicly sharing work
  The same journals we may use to research a problem are often suitable for the submission and publication of SoTL work. Additionally, there are many conference opportunities at the regional, national, or international level that can be great venues for presenting work and obtaining feedback from peers. For instance, there is regularly a SoTL session at the AMS or MAA meetings, and there are SoTL conferences around the world.

5.2.4 Some SoTL Examples

Example 5.1 (Pre-class Reading Assignments) In Chapter 13 of [14], Derek Bruff of Vanderbilt University describes a SoTL project that demonstrates the relationship that often exists between *What works?* questions and *What is?* questions. Bruff describes his "Probability and Statistics for Engineers" course which was taught in an inverted, or "flipped," style, meaning that some of the content delivery moves outside of the class so that during class time, students work on the more challenging course material with the guidance of the expert teacher. In Bruff's case, he moved content delivery outside of class by asking students to read and then respond to some pre-class reading quizzes that contained two or three open-ended questions as well as a "muddiest point" question that asked students to report what they found most confusing or difficult in the reading. After teaching the course, Bruff became interested in whether there were certain types of reading quizzes that would help students better understand the pre-class reading assignments. For instance, perhaps there are certain formats or question types that are more effective at improving understanding of the reading. He realized that in order to answer this question, he needed to know what it is that students were already learning from the pre-class readings. He posed the following as his SoTL question in the *What is?* category:

> In an introduction to statistics course for engineering majors, what are students able to learn by reading their textbooks before class?

Bruff already had data, since he had the open-ended responses from his students to the pre-class reading quizzes. Although he did look at other data, it was the "muddiest point" question that ended up being the most fruitful in this study. He categorized students' muddiest points as being related to factual, conceptual, procedural, or metacognitive knowledge. For instance, he would classify notational problems under the factual category. Conceptual problems were related to probability and statistics ideas. A procedural problem might be related to carrying out a computation or statistical procedure. Metacognition refers to problems in knowing what we do or do not know. He found that, in large, the students had the most problems with concepts. He also found that the types of problems they reported as muddiest points varied by topic. Although this analysis did not directly answer Bruff's above SoTL question, it did answer the related question of what students struggle most with when reading their textbooks. With this information, Bruff was able to move on to study the *What works?* question:

> What kinds of pre-class reading assignments, including questions about the reading, might help students in an introduction to statistics course for engineering majors learn more from reading their textbooks before class?

We won't present that part of Bruff's SoTL study here, but an interested reader can learn more in [14]. Additionally, his work has been presented in SoTL sessions at national conferences. □

Example 5.2 (Do Team Writing Projects Using Real Data Lead to Individual Acquisition of Desired Skills?) In Chapter 12 of [14], John Holcomb of Cleveland State University describes a SoTL project that fits into the category of *What could be?* or a *vision of the possible.* Holcomb was teaching introductory statistics and wanted to show that team data analysis projects could lead to individual competence in data analysis. He developed a series of projects for his course that incorporated real data with the expectation that student groups would analyze the data using software and draw and report appropriate conclusions. In addition, students took individual, take-home midterm and final exams in which they were expected to demonstrate their individual statistics competencies (summarize the data, visualize the data, formulate hypotheses, conjecture results, formally test hypotheses, and write their results in a report [21]). Holcomb describes the evolution of both his question and his data collection plan in [14], but we will present a (simplified) summary here. Since his goal was to demonstrate a *vision of the possible*, i.e., effectiveness of a project-based course, he wasn't comparing results to an existing course model but rather showing that his model was effective. Since the projects in the course were performed in groups, it made sense for him to choose individual take-home exams to see if individual students could complete the data analysis and reporting process. To demonstrate success, he calculated the percentage of students scoring an 80% or better on these examinations. As an additional source of data, Holcomb used student responses from a survey about the writing projects. His surveys contained free-response questions from the qualitative data category. He analyzed this data by "coding" the types of responses and reported finding that 66% of the survey responses were positive. He was able to learn that the students generally found working on the projects in groups to be a positive experience. In addition to reflecting upon his SoTL experience in [14], Holcomb has presented his work at a number of conferences and published a description of the data analysis projects in [22]. □

5.2.5 Benefits of SoTL

There are both personal and professional benefits to engaging in SoTL work. For those that both care about effective teaching and improving the understanding of student learning, SoTL projects can be both interesting and rewarding. On the professional side, research or scholarship is typically part of the expectations of an academic job. Not every university values SoTL work in the same way they might value mathematical research, but there are many that do place value on such scholarship. If you are interested in doing SoTL work as a mathematics faculty member, it is important to know what such work means to your department or institution. Additional professional benefits of doing SoTL work are new connections in the mathematics community and, possibly, new collaborations. By presenting or attending SoTL presentations, I have met many other mathematicians who are interested in problems of teaching and learning and built new professional connections. For some SoTL workers, these connections can lead to collaborations on SoTL projects. Finally, in addition to the individual personal and professional benefits, departments and institutions can benefit from faculty members who work in SoTL. For instance, effective teaching, as studied in SoTL, can improve both student retention rates and the time to graduation [18].

5.2.6 What Background Do You Need to Do SoTL?

If such projects sound interesting to you, you might want to know what you need to do to work in SoTL. If you want to be a mathematician who does work in the Scholarship of Teaching and Learning, then you must first be a mathematician. The importance of content area expertise is discussed in the book *How People Learn*, which was published by the National Research Council [5]. On page 155, the authors state

> Effective teachers need ... knowledge about how to teach in particular disciplines rather than only knowledge of a particular subject matter. ... Expert teachers know the structure of their disciplines, and this knowledge provides them with cognitive roadmaps that guide the assignments they give students, the assessments they use to gauge students' progress, and the questions they ask in the give and take of classroom life.

In addition to content knowledge, it is useful to know some statistics as well as some of the data collection methods of social science research, particularly the qualitative ones as discussed in Sect. 5.2.3. Finally, it is necessary to know the work that has already been performed in the field – or at least to have the resources to find such work, also discussed in Sect. 5.2.3.

5.3 Contributions

5.3.1 Partially Flipped Learning in Linear Algebra

My first SoTL project was related to a Linear Algebra course that I had been planning to teach in spring 2014. Another member of my department at Mines, Deb Carney, and a member of the nearby University of Denver (DU), Nic Ormes, were also scheduled to teach Linear Algebra that semester. We were all interested in trying something new in order to better engage students and decided to work together. Deb and I knew each other well as we worked in the same department, and Nic is Deb's spouse. Deb and I were teaching to a primarily engineering audience at a public engineering university, while Nic was teaching to a broader STEM (Science, Technology, Engineering, and Mathematics) audience at a private liberal arts university. Based upon our own experiences with lecture-based teaching, we instead wanted to use evidenced-based instruction strategies. At that point in time, we read or heard of numerous studies that demonstrated the benefits of active student engagement in the classroom. However, as professors, there can be pressure to teach a large number of topics. One of the issues that can arise in making class time more interactive is that it takes actual time to do so, and we were concerned about how that might affect the number of topics that we would teach. One solution for this seemed to be "flipped learning." The FLIP Learning Network, [46], defines flipped learning as follows:

> **Flipped Learning** is a pedagogical approach in which direct instruction moves from the group learning space to the individual learning space, and the resulting group space is transformed into a dynamic, interactive learning environment where the educator guides students as they apply concepts and engage creatively in the subject matter.

In a flipped classroom, "direct instruction," or knowledge transfer (lecture!), is moved outside of the class meeting time. The method for moving direct instruction outside of class can vary as there are numerous ways to accomplish this goal. For instance, students can be asked to read before class, watch videos before class that were created by the instructor, watch videos before class that were created by others, or some mix of the above. They may also be asked to do some (simple) tasks before attending class. This makes room for class time in which students can work collaboratively on the more challenging portions of the subject matter with the guidance of the faculty expert. This contrasts with the traditional lecture model in which students listen to a lecture as part of a class group and then leave class and work without expert guidance on more challenging tasks. If a flipped class is designed well, the advantage is that students receive help during class time on the parts that are typically harder for them. Any class model can be implemented well or poorly, but we were aware of both research on best practices in designing a flipped classroom and studies that had demonstrated the success of flipped classrooms. This research has been done both in mathematics, (e.g., [34] and [1]) as well as in other fields, such as Physics [13], Economics [26], and Engineering [17]. Additionally, before it became popular at the postsecondary level, flipped learning has been studied for an even longer time at the K-12 level as described in [15]. In 2013, Jacob Bishop and Matthew Verleger analyzed flipped learning education research done at both the K-12 and postsecondary levels and demonstrated both success and best practices [3]. Finally, though it didn't exist at the time, Robert Talbert's book, *Flipped Learning*, [43], is an excellent resource for learning more about flipped classrooms.

5.3.1.1 Our Questions

Although aspects of flipped learning were appealing, we were a bit uncomfortable with the idea of such a drastic change. So we wanted to know if there was some middle ground. We asked, *"Is a partially flipped model effective for student learning?"*. We also asked the related question, *What will our students think about the model?*. We were unaware, but at the time we were planning the course in fall 2013, there was some experimentation around partially flipped classrooms in the K-12 setting as evidenced in an expository article [12]. However, only anecdotal evidence was shared, and a partial flip meant something different than what was described here. In our case, a partial flip meant that we were going to move some of the content outside of the class, incorporate some interactive student engagement during class, but still maintain a small lecture component during class time. We were unfamiliar with the SoTL taxonomy at the time we began our study. That being said, our question lies in the *What could be?* (or *vision of the possible*) category of the SoTL taxonomy. We wanted to show that we could successfully use our course model to teach Linear Algebra. In particular, we chose one or two of the "easier" goals for a lesson to move outside of class into a video that one of us created [7]. Students completed a short set of online questions related to the video content. Class time would begin with an activity that was based upon the pre-class video and would build upon it with the goal of discovering, developing, or previewing the big ideas in the mini-lecture portion that would fill the remaining time.

As an example, we can consider the introduction to linear independence of vectors. The pre-class video contained the following:

1. The definition of a linearly independent set of vectors in $\mathbb{R}^n$
2. A description of a procedure to determine whether a given set of vectors in $\mathbb{R}^n$ is linearly independent
3. Two examples of carrying out the procedure to determine whether or not a collection of vectors in $\mathbb{R}^n$ are linearly independent

Once students arrived to class, they worked in groups to determine when a set containing one vector was linearly independent/dependent and when a set containing two vectors was linearly independent/dependent. Additionally, the lecture time was still somewhat interactive. For instance, during the lecture portion of my class, students were posed some additional true/false questions to test the generalizability of what they determined to be true for one or two vectors. The remaining lecture time was devoted to discussing the general results (e.g., a collection that contains the zero vector is linearly dependent, and a collection of vectors is linearly dependent if and only if one of the vectors in that collection can be written as a linear combination of the others).

5.3.1.2 Results

Due to the fact that most of us were teaching only one section of the course, we chose not to have a control (entirely lecture-based) group. The data we chose to collect was mostly quantitative. We collected data related to student performance such as DFW rates (the percentage of students who receive a D or F in the course or withdraw from the course) and exam performance. We also collected data related to student attitudes by developing a survey with the help of CSM's assessment director. The results of our work were shared at the Joint Mathematics Meetings and a Mathematical Association of America Sectional Meeting and were also published in a special edition of the journal PRIMUS [8] that was devoted to flipped classrooms. Here is a summary of the evidence we collected.

- **DFW Rates**
  First, even though we understood that grades are often quite subjective, universities often look at DFW rates as a measure of how successful students are in a course. We included this information as mostly anecdotal quantitative evidence. At CSM, the cumulative DFW rate for linear algebra over the previous 5 and 1/2 years was 18.6%. That number comes from 25 distinct class sections with 11 distinct instructors. For the partially flipped model, the DFW rate at CSM was 11.3%. At DU, the cumulative DFW rate for linear algebra over the previous 5 and 1/2 years was 22.1%. That number comes from ten distinct class sections taught by seven distinct instructors. For the partially flipped model, the DFW rate at DU was 16.7%. Since I was the only one who had taught the lecture model during that previous 5-year time period, we could draw a more meaningful comparison between the DFW rate for my lecture class and partially flipped class. Before implementing the partially flipped model, the DFW rate for my linear algebra sections was 22.4%. The DFW rate in my partially flipped model was 10%. Again, there are many confounding factors that could potentially explain these differences, and this is an example of one of the challenges that was mentioned in Sect. 5.2.3. We were careful not to conclude that the partially flipped model was the reason for these changes, but we did (cautiously) conclude that the students in the partially flipped model did not appear to be *less* successful than students in the lecture model.
- **Exam Grades**
  The most meaningful student performance comparison came from comparing performance on the final exams in my previously taught lecture-based classes to the partially flipped model. Since final exams are not returned to students, we used the lecture-based final exam with minor modifications as the final exam for the partially flipped group. The average final exam score in the lecture model

Table 5.1 Survey data for 133 students

	Agree	Neutral	Disagree
Watching the videos helps me complete the in-class activities	82%	11%	7%
Watching the videos helps me understand concepts that are new to me	84%	13%	3%
Watching the videos helps me understand concepts that are confusing to me	63%	25%	12%
Watching the videos is good preparation prior to attending class	81%	14%	5%
The format of this class is enjoyable	64%	22%	14%
I recommend that faculty who teach other math courses develop videos for students to view outside of class	55%	31%	15%
I would enroll in another course that uses videos outside of class if I have the opportunity to do so	52%	34%	14%

Table 5.2 Two additional survey questions asked at CSM

	Agree	Neutral	Disagree
The videos helped me learn how to perform basic matrix and vector operations. (For example, row reduce matrices, calculate determinants, and add/multiply matrices)	88.5%	11.5%	0%
The videos helped me recall/restate basic definitions and theorems of linear algebra and/or prove properties of matrices and sets of vectors	45%	25%	30%

with 62 total students was 69.7%. The average final exam score in the partially flipped model was 77%. (This difference was significant at the p=0.0288 level.)

- **Student Attitudes**

  An additional factor that was important to us was student attitudes. We were interested in student perceptions of the experience. Using the abovementioned survey that we developed with the assistance of our university assessment director, we surveyed our students and received 133 responses (out of 159 total enrolled students). Table 5.1 summarizes some of the results.

  Table 5.2 contains questions only asked to students at CSM (93 of the respondents). These questions were asked specifically to CSM students, as they relate to stated course goals in CSM's Linear Algebra class.

  We felt that the data was evidence that our students generally "enjoyed" the partially flipped model. Additionally, we believed that the difference in responses to lines two and three in Table 5.1 was explained, at least in part, by the fact that the videos only contained the "easier" learning goals. For instance, in our linear independence example, we put the definition with basic examples to test a collection of vectors in $\mathbb{R}^n$ for linear independence/dependence. This would likely help someone understand the new definition of linear independence, but maybe it wouldn't do as well in helping someone interpret the meaning of linear independence and its implications, which is often the more confusing part for students. The same could be said for Table 5.2, since such a video might help someone use a calculation to check linear independence, which would fall under line 1 of the table, but such a video might not help someone learn to prove properties of linearly independent or dependent vectors as referred to in line 2 of the table.

At the time of publication, we found the above results sufficiently compelling to share with others, even though we would have liked to include data from additional semesters. However, as the journal in which we published was publishing a special edition devoted to flipped classrooms, it seemed a natural place and time to submit our work. Since then we have continued to adjust and implement our partially flipped model. We have continued to collect evidence, and some of this evidence has been shared at conferences or within other venues. From the data we have analyzed thus far, our results continue to be comparable to what is presented above.

5.3.2 *Supporting Faculty in Pedagogical Change in Calculus II*

Another of my SoTL projects is related to Calculus II. As Mines is a science and engineering university, every student is required to take the three-semester Calculus sequence. We teach Calculus II in approximately 30 sections of roughly 40 students per section over the academic year. My colleague, Deb Carney, and I have served as the primary coordinators for this course through spring of 2022. This means that in addition to teaching the courses, we manage the rest of the instructional team, which consists of a mix of other full-time faculty, adjunct faculty, graduate teaching fellows, and postdocs. We create the common homework and exams that are used across these sections. However, how an individual chooses to instruct a class may vary. In practice, this means that the course goals and assessments (homework and exams) are uniform across the sections, but individual instructors have the freedom to choose how they help students learn Calculus. Deb and I wanted to enable other instructors to choose to make class time more engaging for the students with the hopes that students will be more successful in Calculus II. So our plan was to implement a course model with increased student engagement and to attempt to do so in a way that is easy for others to use. Our course model could also be considered a "partial flip" but was implemented in a slightly different manner. The same research that motivated our Linear Algebra model motivated our Calculus course model, but we had some additional considerations because of the larger multi-instructor setting.

With the support of the Colorado School of Mines Trefny Innovative Instruction Center [44], in 2016, Deb Carney and I were able to participate in an intensive course redesign program in which we developed a course model that had some elements of our partially flipped model for Linear Algebra. The new aspect was the creation of lesson plans that could be used by other instructors that laid out the students' pre-class expectations (video/reading and pre-class questions), suggested beginning-of-class hook problems that would get students started, discussion topics, student activity problems, and discussion/wrap-up suggestions to end class.

The course materials were piloted by Deb and I during the fall semester of 2016 in three sections, and minor adjustments were made. In the spring of 2017, we co-coordinated Calculus II, meaning that we created all of the homework and exams and organized group-grading, common exam times, instructor meetings, etc. A group of 7 faculty members, including Deb and I, taught 13 total sections of Calculus II. We did not require that all faculty members use our course model, but we did offer up the opportunity for others to do so. In the end, three instructors chose not to use the course model and two other instructors, not including Deb and I, decided to try our model. We referred to the seven sections that were using our newly created materials and model as "Flipped" Sections, while the remaining six sections were referred to as "Non-Flipped." Additionally, to differentiate between the sections taught by Deb and I from the other flipped sections, we referred to Deb's sections and my sections as "Flipped-CS" (for Carney and Swanson), while we referred to the others as "Flipped-Other." It should be pointed out that the Non-Flipped sections in some cases incorporated some of our activities and some of the materials but weren't fully committing to the model, while some of the other Non-Flipped sections were taught by instructors who were not necessarily always lecturing but were engaging their students in a variety of other ways. Beginning in Spring 2018, all faculty members teaching Calculus II adopted this model with one exception in Spring 2019. Throughout this time, we not only provided materials but met weekly with faculty to discuss upcoming topics, teaching concerns, and questions.

5.3.2.1 Our Questions

We were interested in a number of questions. The first set of questions addresses student performance across different sections of the course:

- *Were students in the flipped sections at least as successful as the students in the non-flipped sections?*
- *Were there differences in student performance across the three groups (Non-Flipped, Flipped-CS, and Flipped-Other)?*
- *Were there differences in student performance between the Flipped-CS and Flipped-Other sections?*
- *Were there differences in student attitudes between the Flipped-CS and Flipped-Other sections?*

It should be noted that even though we had non-flipped sections, we did not have a true control group. Differences in results could be, at least in part, due to instructor differences or other factors. For this reason, we will continue to adjust our model and gather other types of data in the future.

Our primary question, however, was *"What are the experiences of other faculty members?"*. In [9], we analyze these responses to identify factors that supported pedagogical change. This question lies in the *"What is?"* category, while the student performance questions might be categorized as *"What works?"* questions. We classified our study primarily as a *"What is?"* study and used the student performance as some evidence of success.

5.3.2.2 Results

The data that we collected was both quantitative and qualitative. The qualitative data that we collected was primarily in the form of faculty interviews conducted by collaborators from our teaching and learning center. These were summarized and coded for themes by these collaborators who had the expertise to do this.

We compared individual student performance on certain final exam questions that were tied to course goals related to the themes of Computation, Application, and Reasoning. We made comparisons across all three groups in addition to making comparisons among the flipped groups. Additionally, we collected student surveys to measure attitudes of students in the Flipped sections. These results were presented at the 2018 Joint Mathematics Meetings [10] and are published in [9]. Here is a summary of the evidence we collected.

- Final Exam Questions
  In Calculus II, we had learning goals related to the ideas of Computation, Application, and Reasoning, and the final exam questions were linked back to these themes. We calculated the number of students who scored at or above a 70% on individual problems, as scoring a 70% or better can be viewed as a benchmark of success. Table 5.3 contains a summary of this data. The data is organized into the abovementioned themes, and the "Q" refers to the question number on the exam. The data compares the performance of the non-flipped sections, those that were flipped by Deb Carney and I (Flipped CS) and those that were flipped by other faculty (Flipped Other). The column at the end gives information about the statistical significance of the differences when using a Chi-Square test. The values denoted with (*) were significant at the 0.1 level and values denoted (**) were significant at the 0.01 level. Again, we must be careful to draw conclusions, but we felt that this data demonstrated that the flipped classes were performing at least as well as the non-flipped ones. Additionally, in comparing the two flipped versions, there was only one statistically significant difference (Q2), providing at least some evidence that the other faculty were successfully able to use our course model.
- DFW Rates
  Just as in the previous Linear Algebra study, we can use DFW rates to provide some anecdotal evidence that students are at least not *less* successful in sections applying the partially flipped model than the students in traditional sections (Table 5.4).

Table 5.3 Percentage of students scoring at or above 70%

		Not flipped	Flipped CS	Flipped other	p
Enrollment		172	103	151	
LO1: Computation	Q1	68.6	81.7	74.3	0.06 (*)
	Q2	57.8	77.7	63	0.003 (**)
	Q12	49.4	67.5	59.3	0.009 (**)
LO2: Application	Q3	87.3	95.6	97.2	0.001 (**)
	Q11	53.6	67.3	59.8	0.086 (*)
	Q14	79.5	86.9	82.8	0.261
LO3: Reasoning	Q6	74.7	77.5	71.9	0.617
	Q7	56.5	60.4	64.3	0.065 (*)
	Q8	64.8	82.7	77.4	0.00002 (**)

Table 5.4 DFW rates for Calculus II for the spring semesters 2010–2019

Spring semester 20xx	10	11	12	13	14	15	16	**17**	**18**	**19**
# Sections adopted/total # sections								**8/13**	**13/13**	**11/14**
DFW rate percentage (%)	25	30	21	21	22	14	16	**13.1**	**8.6**	**10.7**

After analyzing both the qualitative and quantitative data, we provided what we believed were the main factors that contributed to the success of the implementation:

- Institutional support for teaching: We were supported by our teaching and learning center in both developing the course model and in collecting data to analyze it.
- The use of evidence-based practices: We developed and implemented a course model that was demonstrated to be effective.
- Creation of faculty community: Many of the other instructors implementing our model discussed how the regular meetings were particularly beneficial as part of their first forays into active learning.

5.4 Future Work and Reflection

Currently, I am working on three other SoTL projects:

- First, the work Deb Carney and I did in [9] occurred before the pandemic began. During the pandemic, we had to transition our course model into both remote and hybrid settings. This also caused us to make some alterations to the model. We are interested in the question *"To what extent did this model successfully translate to a remote/hybrid setting during the pandemic years of 2020-2022?"*. Related to this, we continue to be interested in the experience of other instructors teaching with our model. To this end, we collected survey data from the instructors who taught during the 2020–2022 pandemic years and analyzed the results both quantitatively and qualitatively. This has meant learning how to code qualitative data. This work was recently published [42].

- In addition to our curiosity about the experiences of coordinated faculty teaching Calculus II, I am also interested in the experience of all of our coordinated faculty in the mathematics department at Mines. Our focus is a *"What is?"* question, namely, *"What are the experiences of our coordianted faculty?"*. To this end, Deb Carney and I developed a survey for coordinated faculty with the support of Megan Sanders from our teaching and learning center. This work has been submitted for publication.
- In recent years, I have become frustrated with the relationship between grades and learning, motivating me to join a larger community of faculty members who are developing or implementing less traditional grading schemes. While there is a large spectrum of such schemes, the one that was most appealing to me was something called Mastery Based Testing, which is a form of Mastery Grading [19]. A new postdoc, Aram Bingham, arrived at Mines last fall and was also interested in this system, so we developed our own version of it for Linear Algebra and implemented it in the spring. We are interested in a *"What works?"* question. We have collected and analyzed both qualitative and quantitative data and worked with Carter Moulton from our teaching and learning center to develop a website [29] to display the data as well as the development and implementation process. A paper we wrote related to this work will soon appear in the journal *Teaching and Learning Inquiry*.

All of these projects bring me joy and fulfillment as a mathematics educator, and I will continue to work in SoTL because of what it has meant for me both professionally and personally. First, I have made connections with other mathematicians interested in teaching. This has led to both professional relationships and friendships. Second, working in SoTL has reinvigorated my passion for teaching. I have always enjoyed teaching, but I have a new lens through which to view and approach classroom problems, and this new "SoTL lens" makes approaching such problems exciting. Finally, my work in SoTL has changed how I think about teaching. Instead of viewing an instructor as a "guardian (or gatekeeper) of knowledge," I view the role of the instructor as a "facilitator of learning." This viewpoint influences each and every choice I make when planning a course.

To close, I have presented both an introduction to the Scholarship of Teaching and Learning and some examples to illustrate SoTL, including examples some of the author's SoTL work. For those interested in effective teaching and contributing to the knowledge base about such teaching, SoTL can be an enjoyable and valuable form of scholarship. This has certainly been my own experience!

Acknowledgments First and foremost, I would like to thank Deanna Haunsperger and Steve Kennedy for creating such an impactful program in SMP. I will be forever grateful for my SMP community! Additionally, many thanks to my partners in pedagogy, including Deb Carney, Nic Ormes, Aram Bingham, Megan Sanders, and Carter Moulton. A special thanks go out to Christine Andrews-Larson for her conversations about mathematics education.

References

1. Bagley, S.F.: Improving student success in calculus: a comparison of four college calculus classes. San Diego State University (Doctoral dissertation). Retrieved from ProQuest Dissertations and Theses. (Accession Order No. 3633127) (2014)
2. Bennett, A.G., Manly, I.: Bailout pre-calculus: an approach to improving retention (2018). AMS-MAA Joint Mathematics Meetings, MAA Session on the Scholarship of Teaching and Learning in Collegiate Mathematics
3. Bishop, J., Verleger, M.A.: The flipped classroom: a survey of the research. In: 2013 ASEE Annual Conference & Exposition, vol. 30, pp. 1–18 (2013)
4. Boaler, J.: Mathematical Mindsets: Unleashing Students' Potential Through Creative Math, Inspiring Messages and Innovative Teaching. Jossey-Bass, a Wiley Brand, San Francisco (2015)
5. Bransford, J., Brown, A., Cocking, R.: How People Learn: Brain, Mind, Experience, and School. National Academies Press, Washington, DC (2000)
6. Bromberg, L., Seminelli, M.: The death of paper: should we use digital assessments in undergraduate mathematics? (2018). AMS-MAA Joint Mathematics Meetings, MAA Session on the Scholarship of Teaching and Learning in Collegiate Mathematics
7. Carney, D., Ormes, N., Swanson, R.: COS Linear Algebra Channel. Vimeo. https://vimeo.com/user36100127
8. Carney, D., Ormes, N., Swanson, R.: Partially flipped linear algebra: a team–based approach. Primus **25**(8), 641–654 (2015)
9. Carney, D., Sanders, M., Spiegel, S., Swanson, R., Vasquez, A.: Calculus II course redesign: supporting faculty in pedagogical change. PRIMUS **31**(3-5), 643–657 (2021)
10. Carney, D., Swanson, R.: Flipping calculus II: creating materials that others can use (2018). AMS-MAA Joint Mathematics Meetings, MAA Session on Flipped Classes: Implementation and Evaluation
11. CoMInDS: Mathematical Association of America https://www.maa.org/programs-and-communities/professional-development/cominds
12. Dawes, K., Fox, M.: The partially flipped classroom. Independent Teacher Mag. **11**(1) (2013). https://www.nais.org/magazine/independent-teacher/fall-2013/the-partially-flipped-classroom/
13. Deslauriers, L., Schelew, E., Wieman, C.: Improved learning in a large-enrollment physics class. Science **332**(6031), 862–864 (2011)
14. Dewar, C.B..J.: Doing the Scholarship of Teaching And Learning in Mathematics. The Mathematical Association of America (2015)
15. Finkel, E.: Flipping the script in k12. District Administration **48**(10), 28 (2012). https://www.districtadministration.com/article/flipping-script-k12
16. Freeman, S., Eddy, S.L., McDonough, M., Smith, M.K., Okoroafor, N., Jordt, H., Wenderoth, M.P.: Active learning increases student performance in science, engineering, and mathematics. Proc. Natl. Acad. Sci. **111**(23), 8410–8415 (2014)
17. Gannod, G.C., Burge, J.E., Helmick, M.T.: Using the inverted classroom to teach software engineering. In: Proceedings of the 30th International Conference on Software Engineering, pp. 777–786 (2008)
18. Gyurko, J., MacCormack, P., Bless, M.M., Jodl, J.: Why Colleges and Universities Need to Invest in Quality Teaching More Than Ever: Faculty Development, Evidence-Based Teaching Practices, and Student Success. Association of College and University Educators, New York (2016). http://acue.org/quality-teaching/
19. Harsy, A.: Variations in mastery-based testing. Primus **30**(8–10), 849–868 (2020)
20. Hartlage, B., Higginbotham, D., Pierce, I., Schaubroeck, B.: Engineering learning in calculus at the colorado school of mines (2018). AMS-MAA Joint Mathematics Meetings, MAA Session on the Scholarship of Teaching and Learning in Collegiate Mathematics

21. Holcomb, J.: Assessing student learning in introductory statistics: an authentic assessment approach. In: ARTIST Roundtable Conference on Assessment in Statistics. Lawrence University (2004)
22. Holcomb Jr, J.P., Ruffer Jr., R.L.: Using a term-long project sequence in introductory statistics. Am. Stat. **54**(1), 49–53 (2000)
23. Hutchings, P.: Approaching the scholarship of teaching and learning. In: Opening lines: Approaches to the scholarship of teaching and learning, pp. 1–10. Carnegie Foundation for the Advancement of Teaching, Menlo Park (2000)
24. Klamerus, L., Carlson, C., Harsy, A.: Analyzing the impact of mastery-based testing in mathematics courses (2018). AMS-MAA Joint Mathematics Meetings, MAA Session on the Scholarship of Teaching and Learning in Collegiate Mathematics
25. Kornelson, K.: Course innovation: a large-scale active learning program in pre-calculus and trigonometry (2018). AMS-MAA Joint Mathematics Meetings, MAA Session on the Scholarship of Teaching and Learning in Collegiate Mathematics
26. Lage, M.J., Platt, G.J., Treglia, M.: Inverting the classroom: A gateway to creating an inclusive learning environment. J. Econ. Educ. **31**(1), 30–43 (2000)
27. Lester, F.K.: On the theoretical, conceptual, and philosophical foundations for research in mathematics education. ZDM Math. Educ. **37**(6), 457–467 (2005)
28. Lynch, A.: Peer problem review in calculus I (2018). AMS-MAA Joint Mathematics Meetings, MAA Session on the Scholarship of Teaching and Learning in Collegiate Mathematics
29. Mastery Based Testing in Linear Algebra: Trefny Center, Colorado School of Mines. https://trefnycenter.mines.edu/resources/masterygrading/
30. Miller-Young, J., Yeo, M.: Conceptualizing and communicating SoTL: a framework for the field. Teach. Learn. Inquiry **3**(2), 37–53 (2015)
31. Nair-Hart, G.: Inquiries into student understanding and insights into student misunderstanding (2018). AMS-MAA Joint Mathematics Meetings, MAA Session on the Scholarship of Teaching and Learning in Collegiate Mathematics
32. Nelson, C.: Doing it: Examples of several of the different genres of the scholarship of teaching and learning. J. Excellence College Teaching **14**(2), 85–94 (2003)
33. Onwuegbuzie, A.J., Leech, N.L.: On becoming a pragmatic researcher: the importance of combining quantitative and qualitative research methodologies. Int. J. Soc. Res. Methodol. **8**(5), 375–387 (2005)
34. Overmyer, G.: The Flipped Classroom Model for College Algebra: Effects on Student Achievement. Doctoral Dissertation (2014)
35. Patel, V.: Training Graduate Students to Be Effective Teachers. Chronicle of Higher Education (2017)
36. Potter, M.K., Kustra, E.D.: The relationship between scholarly teaching and SoTL: models, distinctions, and clarifications. Int. J. Scholarsh. Teach. Learn. **5**(1), 23 (2011)
37. PRIMUS: Problems, Resources, and Issues in Mathematics Undergraduate Studies. https://www.tandfonline.com/loi/upri20
38. Project NExT (New Experiences in Teaching), Mathematical Association of America. https://www.maa.org/programs-and-communities/professional-development/project-next
39. Rivers, Y.: How is growth mindset and metacognitive self awareness linked in developmental math students? (2018). AMS-MAA Joint Mathematics Meetings, MAA Session on the Scholarship of Teaching and Learning in Collegiate Mathematics
40. Schoenfeld, A.: Research in mathematics education. Rev. Res. Educ. **40**(1), 497–528 (2016)
41. Sterrett, A.: 101 Careers in Mathematics. Mathematical Association of America, Washington, DC (2014)
42. Swanson, R., Carney, D., Sanders, M.: Adapting and Sustaining Change: A Case Study of a Successful Calculus II Course Redesign during the Pandemic Years. PRIMUS, pp. 1–14 (2024). https://doi.org/10.1080/10511970.2024.2352364
43. Talbert, R.: Flipped Learning. Stylus Publishing LLC, Sterling (2017)
44. Trefny Center: Colorado School of Mines. https://trefnycenter.mines.edu

45. We Use Math. http://www.weusemath.org
46. What is flipped learning?: Flip Learning Network. https://flippedlearning.org/definition-of-flipped-learning/
47. Wijetunge, T., Premadasa, K.: How do students select questions in a math exam? (2018). AMS-MAA Joint Mathematics Meetings, MAA Session on the Scholarship of Teaching and Learning in Collegiate Mathematics

Chapter 6
My Journey in Geometry

Elizabeth Stanhope

Abstract My scholarly work to date falls into three categories: spectral geometry, spectral graph theory, and discipline-based education research. Spectral geometry is the oldest and most developed branch of my work and so it is the focus of this article. After a brief discussion of the origin of spectral geometry, Sect. 6.2 carefully considers whether or not we can hear the length of a vibrating string. This problem provides a concrete example of the fundamental objects in classical spectral geometry: the Laplace operator, its eigenvalues, and eigenfunctions. Using the vibrating string as a stepping stone, the discussion moves to the spectrum of the Laplace operator on a Riemannian manifold. Questions asked by spectral geometers in this setting are described and two important results about the Laplace spectrum of a manifold are discussed. Section 6.3 introduces the Steklov operator and its spectrum, showcasing a few results concerning the Steklov spectral geometry of a manifold with boundary. The objects that are the focus of my work in spectral geometry, called orbifolds, are described in Sect. 6.4.1. After that my contributions to the study of the Laplace and Steklov geometry of orbifolds are discussed. Sections 6.2, 6.3, and 6.4.1 are written to be accessible to a student who is familiar with differential equations, multivariable calculus, and linear algebra, respectively. Sections 6.4.2 and 6.4.3 will be easier to read for those with some background in abstract algebra and point-set topology, respectively. The article concludes with a brief discussion of my work in spectral graph theory and education research, as well as some thoughts about the next steps in my research program.

Keywords Orbifold · Geometry · Eigenvalue · Matrix · Differential equation · Manifold · Sound · lgbt · Graph theory · Math education · Spectrum · Symmetry · Topology

---

E. Stanhope (✉)
Lewis & Clark College, Portland, OR, USA
e-mail: stanhope@lclark.edu

A. Brisbin et al. (eds.), *Research Connections*, Association for Women in Mathematics Series 36, https://doi.org/10.1007/978-3-031-81217-0_6

6.1 Career Trajectory

It was in the summer of 1987 that math moved from a just a peripheral interest to front and center in my life. I was at an eighth-grade math and technology summer camp. I had learned just enough about computer programming and just enough about complex numbers to write a program to generate a Julia set. As the grayscale image of the Julia set tapped itself out onto a spool of computer paper, I was mesmerized by the fact that math could produce an image that so well mimicked patterns in nature. The fractal images of the eighties hinted to me that math was much deeper than solving for x and rise-over-run.

My curiosity about math had been nurtured from an early age. When I was little I devoured my copies of *GAMES Magazine* and my mom regularly told me that I "would love algebra because it is all about puzzles." And it was true! My eighth-grade algebra teacher was fantastic; she always had a fun problem to puzzle over. My brain was further stretched by high school math teachers who would say things like "The human brain can process information three-times faster than typical speech, so I will teach this class at triple speed." I will never forget the image of my calculus teacher walking across the classroom to place fifty cents on my desk (enough to buy a soda) when I was the only one to solve the "man walking away from a lamppost"-related rate problem.

As a seventeen-year-old I decided that it was boring to know your college major before you arrived at college. So despite my motivation to study math, in my first two years at Carleton College (1991–1993), I tried out being an art, geology, Japanese, and religion major. Despite my enjoyment of all of these topics, I felt that the objects and proofs in my math classes were far more captivating. My eighth-grade obsession with fractals found a home in point-set topology when Professor Gail Nelson defined the Cantor set. I could not believe this set was simultaneously uncountable and had measure zero.

Being a math major was not easy; my community on campus was disjoint from the college's math community. While I was successful with increasingly complex mathematical ideas during my teenage years, I struggled to keep up with the growing complexity of my place in the world. I was a feminist during the height of "New Conservatism" and a closeted queer person from small town USA. My college friends were reading Simone de Beauvoir and creating queer performance art, while I was holed up in a study room teasing out the relationship between field extensions and Galois groups. These two parts of my life were discouragingly distant from each other.

My summer jobs during college were testing grounds for possible career directions. After my first year of college I worked for Baltimore County Recreation and Parks, a world I had grown up in because my father was the County's Chief Naturalist. I discovered a passion for teaching as I led nature education programs for county summer camps. The next summer I was a teaching assistant for the Johns Hopkins University Center for Talented Youth. I supported a group of middle school students as they learned problem-solving and proof writing. This opportunity to apply my teaching skills in a mathematical setting reassured me that a career in teaching math was a good fit. During my junior and senior summers I was a research intern at the Geometry Center at the University of Minnesota. The center was best described as a mathematical Disneyland. I was immersed in a world of tori and Klein bottles, huge models depicting three-dimensional shadows of four-dimensional polytopes and state-of-the-art animations of differential geometry. Mathematics transformed from problem sets and textbooks into a living world of inquiry and exploration, a world I wanted to get back to after college.

As my confidence grew about pursuing a career in mathematics, mathematical peer and mentoring relationships became vitally important. Summer 1994 at the Geometry Center and fall 1994 with the Budapest Semesters in Mathematics Program introduced me to friends who both enjoyed math and had connections to my nonmathematical interests. I grew to enjoy doing mathematics as part of a team of friends. My sense of identity as a mathematician was further solidified when my Geometry Center mentor, Don Aronson, introduced me to a group of visitors as a mathematics "pre-doc." Don's comment was brief but had a lasting impact. When I returned to my home campus after the Budapest Program I met the Math Department's new hires, Deanna Haunsperger and Steve Kennedy. Their invitation to be a teaching assistant for the new "Carleton/St. Olaf Summer Math Program for Women (SMP)" was totally unexpected and incredibly exciting. In this program I was able to be my whole self while supporting program participants as they navigated their own experiences as women in mathematics.[1]

After teaching math for a few years after college, my desire to teach and do mathematics beyond the high school level drove me to apply to graduate school. Fortuitously, my college major advisor Mark Krusemeyer suggested I apply to Dartmouth College. The Department of Mathematics at Dartmouth built a supportive graduate environment by recruiting gender-balanced cohorts of graduate students and hosting social events that connected students and faculty. I completed my dissertation under the guidance of geometer Carolyn Gordon alongside her other advisees at that time, Emily Dryden and Emily Proctor. I was never the only woman in the room! In addition to her amazing mathematical mentorship, Carolyn's explicit advice not to compartmentalize my life and ability to name issues such as imposter syndrome were invaluable. A research group potluck where visiting international researchers, kids playing croquet, students, faculty, and family members all came together exemplified the open culture that was so important to my PhD experience.

While I was applying for jobs during my final year at Dartmouth, then post-doc Craig Sutton encouraged me to apply for a position that felt like a big reach: a post-doc at the University of Michigan. What might have seemed like a small bit of encouragement to Craig had a huge impact on my career path. I was offered the post-doc and moved to Michigan in the summer of 2002. Teaching in the

[1] My current students use the phrase "gender minorities in mathematics" rather than "women in mathematics" in order to include people with non-binary gender identities. I prefer the non-binary language but do not use it here as it was not the language used at the time.

Michigan calculus program, led at the time by another wonderful mentor Karen Rhea, deepened my teaching skills. Topologist Richard Canary reached out and pulled me into his community of PhD students and post-docs. I also met geometer Alejandro Uribe and we began a collaborative project that would later be supported by an Association for Women in Mathematics Mentoring Travel Grant. Although I was at Michigan for only a year, the experience there deepened my mathematical skill set and connections tremendously.

As my career progressed, my connection to SMP regained importance. In returning to the program as part of the SMPosium in 2005 and 2007, I found a community filled with valuable advice about the tenure-track, as well as support during the times when I felt isolated in my work. I was delighted to return to SMP in 2012 as an instructor, closing the loop on my SMP experience by becoming a mentor for a new cohort of participants. Inspired by SMP and the incredible mentors I have had the privilege of working with, my current work at Lewis & Clark College seeks to make math and other STEM careers more accessible to all. The graph theory and BioSQuaRE projects described below are motivated by this goal. More recently, I have co-led two research workshops titled "Women in Geometry." The first was hosted by the Banff International Research Station in Canada in 2015. The second was hosted in 2019 by the Casa Matemática Oaxaca in Mexico. These workshops aim to build collaborative research networks between women doing research in geometry.

Essential to nearly the entire story above has been my relationship with my partner Keska Kemper. She was the first to show me the way to be a math/science enthusiast, feminist, and member of the queer community. Together we have negotiated the twists and turns of first jobs, graduate school, establishing careers, and living those careers while raising two wonderful kids. Her presence is woven into every aspect my life.

6.2 Can You Hear the Shape of a Drum?

The story of the origin of spectral geometry is beautifully told by Carolyn Gordon and David Webb in the Chauvenet Prize[2] winning article [13]. The background provided below follows the beginning of [13] while going into less detail.

In the nineteenth century astronomers discovered that they could use the spectrum of light emitted by a star to determine the chemical elements present in the star. This is because each gas in a star, such as hydrogen or helium, absorbs a discrete set of light frequencies that form a distinctive signature of that gas. The image in Fig. 6.1 shows the spectrum of visible light present in sunlight. The dark bands in this spectrum, called Fraunhofer lines, indicate missing light frequencies.

[2] This prize is awarded by the Mathematical Association of America to recognize the year's top expository writing in mathematics.

Fig. 6.1 Solar spectrum [28]

Missing frequencies match up to the signature light frequencies of gases present in the sun. In fact helium is named after the Greek sun god Helios because the signature color spectrum of helium was first detected on the sun [16].

Motivated by this work, in 1882 British physicist Sir Arthur Schuster offered the following observation [13]:

> We know a great deal more about the forces which produce the vibrations of sound than about those which produce the vibrations of light. ... but it would baffle the most skillful mathematician to solve the inverse problem and to find out the shape of a bell by means of the sounds which it is capable of sending out.

The question of determining the shape of an object given only the object's intrinsic harmonic properties has blossomed into a rich field of mathematical study known as inverse spectral geometry. The infinite list of fundamental vibrational frequencies of an object is known to mathematicians as the *Laplace spectrum* of the object.

6.2.1 Can You Hear the Length of a String?

A first inverse spectral geometry question to consider is whether or not we can hear the length of a string. View the straight line labeled (a) in Fig. 6.2 as a string on a harp. When plucked, the string can vibrate in a variety of ways including the standing waves illustrated in (b), (c), and (d). The frequency of waveform (b) is the lowest one at which the string can vibrate and corresponds to the pitch of the string. Waveforms (c) and (d) vibrate at increasingly higher frequencies called the first and second overtones of the string, respectively. Dividing the string into bumpier and bumpier oscillations yields an infinite list of standing waves which vibrate at higher and higher frequencies.

These frequencies can be computed by modeling the vibrating string using methods from partial differential equations. First, assume the string is stretched out along the x-axis in the Cartesian plane from $x = 0$ to $x = L$. Note that we are taking L to be the length of the string. We seek a function $u(t, x)$ that gives the vertical displacement of the string over point x, for $0 \leq x \leq L$, at time t, for $t \geq 0$.

Fig. 6.2 Three fundamental oscillations of a string

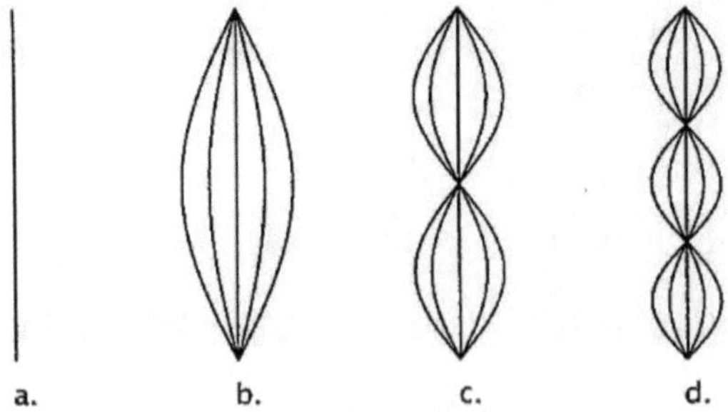

To model vibration we require $u(t, x)$ to satisfy the wave equation:

$$c^2 \frac{\partial^2 u}{\partial x^2} = \frac{\partial^2 u}{\partial t^2}.$$

This equation is derived by modeling the kinetic and potential energy of the string and applying the physical law of conservation of energy. The fact that the ends of the string do not move is captured by requiring that for all t, $u(t, 0) = u(t, L) = 0$. For simplicity we will take constant $c = 1$.

Applying the method of separation of variables,[3] we look for solutions u of the form $u(t, x) = A(t)\psi(x)$. This reduces the problem to solving a pair of ordinary differential equations:

$$\frac{d^2 A}{dt^2} = -\lambda A \qquad \text{(time differential equation)} \tag{6.1}$$

$$\frac{d^2 \psi}{dx^2} = -\lambda \psi \qquad \text{(space differential equation)} \tag{6.2}$$

for some nonnegative constant λ.

But wait, something very exciting has just occurred! Before we use this reduced problem to compute the fundamental vibrational frequencies of our string, let us take a closer look at the space differential equation above. Observe that the process of computing the second derivative of a function is a linear transformation. In fact it is a famous linear transformation: The second derivative operator $\frac{d^2}{dx^2}$ is the one-dimensional Laplace operator on our interval $[0, L]$. It is the cousin of the three-dimensional Laplace operator $\frac{\partial^2}{\partial x^2} + \frac{\partial^2}{\partial y^2} + \frac{\partial^2}{\partial z^2}$ that appears in multivariable calculus courses. Let the symbol Δ denote the Laplace operator in any context. Now observe that the space differential equation (6.2) asks us to find functions ψ for which the image of ψ under the Laplace operator is simply a scaled copy of ψ

[3] To see this process in full detail see Chapter 10 in [1], for example.

back again. We have happened upon the eigenvalue problem $\Delta\psi = -\lambda\psi$. This is exciting because eigenvalue problems are the heart of all of my work. In this section we are considering the eigenvalue problem for the Laplace operator, in Sect. 6.3 we study the eigenvalue problem for the Dirichlet-to-Neumann operator, and Sect. 6.5.1 is built around the eigenvalue problem of the adjacency matrix of a graph.

Now let us get back to the task at hand. Methods from differential equations can be used to show that solutions to Eqs. (6.2) and (6.1) have the form

$$A(t) = a\cos(\sqrt{\lambda}t) + b\sin(\sqrt{\lambda}t)$$

$$\psi(x) = c\cos(\sqrt{\lambda}x) + d\sin(\sqrt{\lambda}x)\,,$$

respectively, where a, b, c, and d are arbitrary real constants. Applying the condition that $u(t, 0) = u(t, L) = 0$ to solutions to the space differential equation forces c to be zero and $\sqrt{\lambda}L = n\pi$ for n a nonnegative integer.

Assembling these facts back into the form $u(t, x) = A(t)\psi(x)$ yields a family of solution functions:

$$u(t, x) = \left(a\cos\left(\frac{n\pi}{L}t\right) + b\sin\left(\frac{n\pi}{L}t\right)\right)\sin\left(\frac{n\pi}{L}x\right).$$

For a fixed value of t we see that solutions are a scaled sine wave of period $\frac{2L}{n}$ that is zero at the endpoints of the string. Note that n counts the number of "bumps" in the wave as in Fig. 6.2 where wave (b) has one bump, (c) has two bumps, and (d) has three bumps. We obtain a standing wave solution to the wave equation that oscillates with frequency $\frac{n}{2L}$ as we let t vary.

We can now answer the question of whether or not we can hear the length of a string. The list of fundamental vibrational frequencies (also called harmonics) of the string is $\frac{1}{2L}, \frac{2}{2L}, \frac{3}{2L}, \ldots, \frac{n}{2L}, \ldots$. If we take the reciprocal of the lowest frequency on this list (the pitch of the string) and divide the result by two, we recover the length L. So yes, we can hear the length of a string.

6.2.2 *Listening to Higher-Dimensional Objects*

In spectral geometry we examine the harmonic properties of objects more general than the interval $[0, L]$ in the real line. An important class of objects studied by spectral geometers is Riemannian manifolds. A Riemannian manifold is a metric space to which the tools of calculus apply. Figure 6.3 illustrates three Riemannian manifolds, from left to right: the sphere, the hemisphere, and the torus. Each object is two-dimensional because we consider only the surface of the object, not its inside. For example, the torus is not solid like a donut, but instead is like an inner tube whose rubber skin is infinitely thin. Each of the surfaces in Fig. 6.3 bends smoothly and has no sharp corners. At a sharp corner, the tools of calculus break down and so

Fig. 6.3 Sphere, hemisphere, and torus

sharp corners are not allowed on a manifold. First described by Riemann in 1854, Riemannian manifolds have proven to be extremely useful objects. They provide the mathematical language needed to express Einstein's theory of general relativity and now appear frequently in many areas of mathematical physics.

Viewing a manifold as a wildly shaped drum, we ask "Can you hear the shape of a manifold?" To make this question mathematically precise we need to return to the vibrating string for a moment. Recall that the space differential equation (6.2) can be viewed as the eigenvalue problem $\Delta\psi = -\lambda\psi$ where Δ is the one-dimensional Laplace operator on the interval. The fact that the string has infinite list of harmonics $\frac{1}{2L}, \frac{2}{2L}, \frac{3}{2L}, \ldots, \frac{n}{2L}, \ldots$ is equivalent to the fact that there is an infinite list of eigenvalues λ_n that satisfy the eigenvalue problem $\Delta\psi = -\lambda\psi$. The connection is that the nth vibrational frequency is equal to $\frac{\sqrt{\lambda_n}}{2\pi}$. Using the tools of Riemannian geometry, one can define a Laplace operator on any Riemannian manifold that generalizes the familiar Laplace operator on $\mathbb{R}^n$. It can be shown that the eigenvalues of the Laplace operator on a compact Riemannian manifold will always form a sequence of nonnegative numbers $\lambda_1 \leq \lambda_2 \leq \lambda_3 \leq \cdots \to +\infty$ where each eigenvalue can be repeated only finitely many times. The list of eigenvalues is called the Laplace spectrum of the manifold. When a spectral geometer asks, "Can you hear the shape of a manifold?" they are really asking, "Given the Laplace spectrum of a Riemannian manifold, what can be said about the geometry or topology of the manifold?"

We have seen that the Laplace spectrum of a string determines the shape of the string as we can determine its length from the lowest eigenvalue. This might lead us to believe that the Laplace spectrum of a manifold determines the shape of the manifold; however, this is not the case. It is possible for two differently shaped manifolds to have identical Laplace spectra. Such manifolds are said to be *Laplace isospectral*. The first example of a pair of manifolds with this property was given by John Milnor in 1964 in a one-page paper in the Proceedings of the National Academy of Science [15, 21]. Milnor constructed isospectral (having the same Laplace spectrum), non-isometric (having different shapes) sixteen-dimensional tori. This remarkable result weaves together number theory (modular forms), algebra (Leech lattices), and analysis (the Poisson summation formula).

Milnor's foundational result opened the floodgates to questions in spectral geometry. Which features of a manifold are determined by its Laplace spectrum? Which features are not? How many different manifolds can share the same Laplace

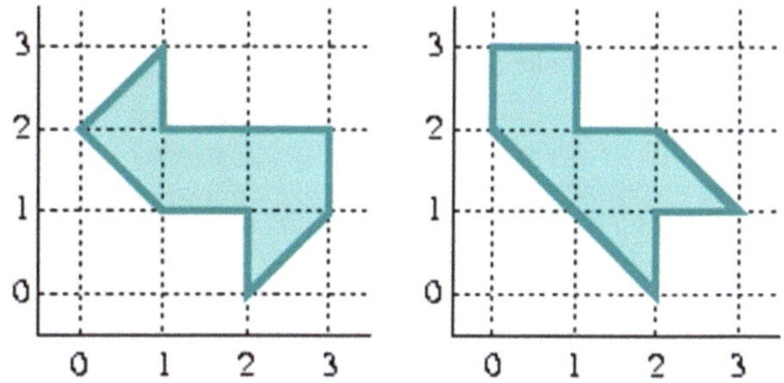

Fig. 6.4 Laplace isospectral domains that are not congruent [8]

spectrum? Can a manifold be deformed in a way that leaves its Laplace spectrum unchanged? Many results have been obtained over the last fifty years and many questions are still being investigated. In 1992 Gordon et al. [14] were the first to construct "drums that sound the same." That is, they discovered a pair of planar domains that are noncongruent but share exactly the same Laplace spectrum. Pairs of planar domains with this property are often called "isospectral drums." Interestingly, it can be shown that the Laplace spectrum does determine the area and the perimeter of a planar domain. The drums found by Gordon, Webb, and Wolpert are shown in Fig. 6.4. By counting triangles you can check that the perimeter and area agree for these two shapes.

6.3 From Drums to Imaging

Twenty years after Sir Arthur Schuster asked if one could detect the shape of a bell from the sound of its ringing, Russian mathematician Vladimir Steklov introduced a problem in 1902 that has recently caught the interest of spectral geometers [12, 19]. An applied interpretation of the problem is to seek a nondestructive way to detect cracks in the interior of a bridge support. The imaging method called "electrical impedance tomography [EIT]" provides a way to scan the interior of the support from its exterior [18]. In EIT electrodes are placed in a circle around a cross section of the bridge support. A known amount of electrical current is then fed into the column. This current on the surface of the column induces an electrical potential field that crosses through its interior. The magnitude of this induced field on the surface of the column is then measured by the circle of electrodes. In short: a current is input around the edge of a cross section, it travels through the column, and the resulting output current around the edge of the cross section is recorded. Because cracks on the interior of the column affect its conductivity, the way the input current transforms as it passes through the column can be used to construct a scan of the cross section of the column. An example is given in Fig. 6.5.

The use of an electrical current to scan the interior of a column appears to be completely different from the problem of detecting the shape of an object from the way it resonates. What makes EIT a question in spectral geometry? The key is to

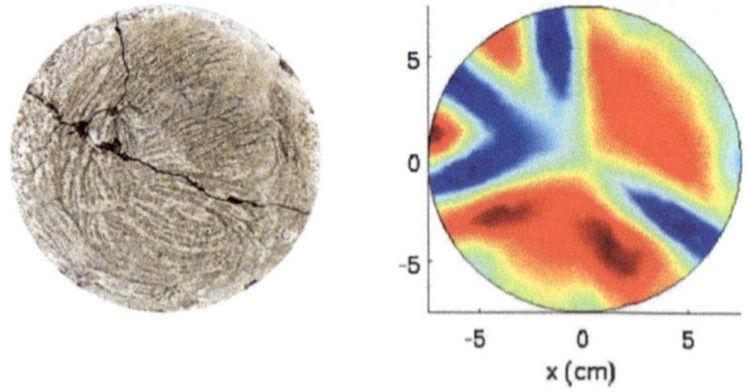

Fig. 6.5 An EIT scan of a short concrete cylinder [24]

remember that the heart of a question in spectral geometry is always an eigenvalue problem. To relate EIT to spectral geometry we need to rephrase EIT in terms of a linear operator and its eigenvalue spectrum.

To do this first observe that we have a mapping between functions defined on the boundary of a cross section of the column. The input current is transformed into the output current. The problem posed by Steklov examines this process in the mathematical setting of manifolds *with boundary*. The hemisphere in Fig. 6.3 is an example of a manifold with boundary. If you trace the surface of the sphere or torus in Fig. 6.3 with your finger, you will forever loop around and around the surface. However, if you trace along the surface of the hemisphere, you eventually arrive at the circular edge at the top of the hemisphere and have to stop. This circular edge is called the boundary of the hemisphere. Both the sphere and the torus have no boundary. Suppose M is an n-dimensional manifold with boundary. Then the boundary of M is an $(n-1)$-dimensional manifold often denoted by ∂M. The Steklov problem takes a function f defined on ∂M, extends it to a unique function $\tilde{f}$ on all of M for which $\Delta \tilde{f} = 0$, and returns a function $\mathcal{D}f$ defined on ∂M that records how quickly $\tilde{f}$ is changing near ∂M. (More precisely, $\mathcal{D}f$ is the *normal derivative* of $\tilde{f}$ on ∂M.) The fact that $\Delta \tilde{f} = 0$ on M means that $\tilde{f}$ is a good model of the distribution of electrical charge over the interior of M. The operator $\mathcal{D}$, mapping f to $\mathcal{D}f$, is called the Dirichlet-to-Neumann operator. Just like the Laplace operator, the Dirichlet-to-Neumann operator has a list of nonnegative eigenvalues written $0 \leq \sigma_1 \leq \sigma_2 \leq \sigma_3 \leq \cdots \to +\infty$. This list is called the *Steklov spectrum* of a manifold with boundary.

The Steklov spectrum of a disk of radius R is relatively easy to compute because the eigenvalue problem $\mathcal{D}f = \sigma f$ on the disk is solved by a particularly nice set of functions: the spherical harmonics. For an index $k = 0, 1, 2, \ldots$ the degree k spherical harmonic functions are all functions $c_1 r^k \cos(k\theta) + c_2 r^k \sin(k\theta)$ where c_1 and c_2 vary through the real numbers, and $0 \leq r \leq R, 0 \leq \theta \leq 2\pi$ parametrize the disk. Figure 6.6 shows the graph of a degree eight spherical harmonic on the unit disk. The function strives to be zero on the interior of the disk reflecting the fact that the Laplacian of the function must be zero there. At the boundary it smoothly bends into a sinusoidal function, dividing the boundary circle into bumps similar to the standing waves on the string in Fig. 6.2. The functions $r^k \cos(k\theta)$ and $r^k \sin(k\theta)$

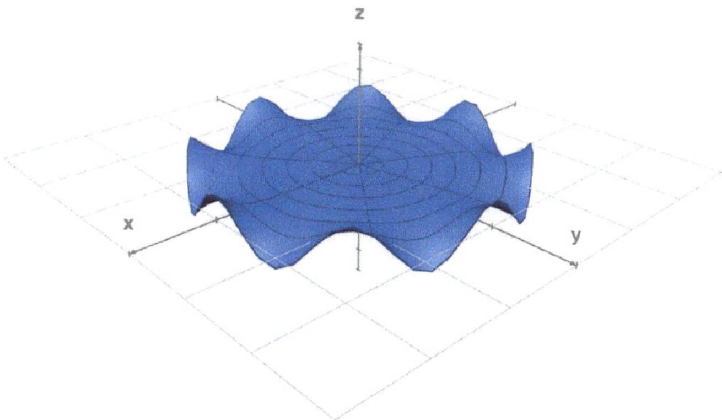

Fig. 6.6 Degree eight spherical harmonic on the unit disk

each contribute a copy of the eigenvalue k/R to the Steklov spectrum of the disk. Thus, the Steklov spectrum of this disk is $0, \frac{1}{R}, \frac{1}{R}, \frac{2}{R}, \frac{2}{R}, \frac{3}{R}, \frac{3}{R}, \ldots$.

We can obtain information about the shape of a manifold with boundary from its Steklov spectrum. Examples of Steklov isospectral manifolds of dimension three and greater [6, 12] show that the Steklov spectrum of a manifold does not determine its shape. However, in contrast to the Laplace spectrum setting, it is not yet known if the Steklov spectrum determines the shape of a planar domain. Excellent surveys of Steklov spectral results are presented in [7, 12].

6.4 My Contributions to Spectral Geometry

6.4.1 Symmetry and Orbifolds

My research focuses on objects called Riemannian orbifolds. To understand what these are we begin by examining the symmetries of the beautiful pattern below that was found in an Egyptian tomb[4] and reproduced by Owen Jones in the text [17].

In analyzing the symmetries of this pattern, imagine that the square in Fig. 6.7 has been repeated to tile all of the plane. The symmetries we seek are the ones that shift the plane and bring the pattern right back to itself. Figure 6.8a illustrates three such symmetries. We can slide the plane to the right and move the central yellow dot over to the center of the adjacent flower. In this way we move the plane without causing the pattern to become visibly offset from its original position. Figure 6.8a also shows that we can slide the plane vertically to move the central yellow dot to

[4] Having used this image in numerous talks I have wondered: "Who painted this image?" "Whose tomb was it?" A brief exploration of Jones' text led to more questions than answers.

Fig. 6.7 Pattern from an Egyptian tomb [17]

(a) (b) (c)

Fig. 6.8 Obtaining the torus from translations. (**a**) Three translations. (**b**) Orbit of central dot. (**c**) Fundamental domain for a torus

the center of the flower immediately above it, or slide the plane to the left by two "flower units" without affecting how the pattern spreads across the plane. All three of these symmetries are called translations; they can be achieved algebraically by choosing a fixed vector in $\mathbb{R}^2$ and adding that vector to every point in the plane. We will let $\mathcal{T}$ denote the set of translational symmetries of this pattern.

Now suppose we slide the central yellow dot around the plane using every single symmetry in $\mathcal{T}$, each time recording its position with a new yellow dot. We end up with an infinite collection of yellow dots spreading across the plane in a grid-like pattern, as illustrated in Fig. 6.8b. This infinite set of dots is called the *orbit* of the central yellow dot under the action of $\mathcal{T}$ on the plane. The black dots in Fig. 6.8c are another orbit of the action of $\mathcal{T}$ on the plane. One can show that the property of two points being in the same orbit is an equivalence relation, and so the set of orbits of all of the points in the plane is a partition of the plane.

So what is the rectangle in Fig. 6.8c? That rectangle cuts out a portion of the plane that contains one and only one point from every orbit of the action of $\mathcal{T}$ on the plane. But there is a problem! The translation that shifts the plane to the right one "flower unit" moves points on the left side of this rectangle to points on the right side of the rectangle. This is a problem because the rectangle is not allowed to contain two points from the same orbit. To fix this we glue these two sides together, the gluing indicated by the double arrows on each of these vertical sides. Similarly

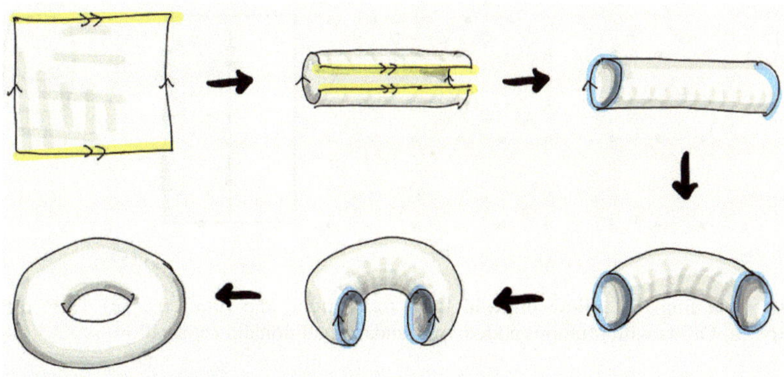

Fig. 6.9 Stretching and gluing a rectangle into a torus

the top and bottom of the rectangle are also glued, indicated by the single arrows on these sides. This rectangle is the *fundamental domain* of the action of $\mathcal{T}$ on the plane.

Now let us step back for a moment. A rectangle with sides glued as indicated in Fig. 6.8c might just be a familiar sight. If you imagine the rectangle is made out of a stretchy material and perform the gluings as illustrated in Fig. 6.9, the resulting object is another torus like the one in Fig. 6.3. The procedure of analyzing the orbits of points in the plane under the action of $\mathcal{T}$, identifying a fundamental domain for this action, and obtaining a surface by gluing the fundamental domain according to the arrows in its edges is called *taking the quotient of the plane by the symmetries in* $\mathcal{T}$. In taking this quotient we go from a set[5] of symmetries acting on the plane to an interesting object – a manifold, which in this case is a torus.

We now modify the quotient process above slightly and in doing so make the connection to orbifolds. To do this first return to the pattern in Fig. 6.7 and observe that there are symmetries we have not yet discussed. They are the 180° rotations such as the one illustrated in Fig. 6.10a. We expand our set of symmetries from just the translations in $\mathcal{T}$ to translations, rotations, and all of the symmetries you get by combining translations and rotations. Denote this new, bigger set of symmetries by $\mathcal{S}$. Figure 6.10b shows two orbits of the action of $\mathcal{S}$ on the plane. Although the orbit of the central yellow dot has not changed, the orbit indicated by the black dots has grown. Our old torus fundamental domain from Fig. 6.8c is shown again in Fig. 6.10b. It now contains two black dots and so fails to be a fundamental domain for the action of $\mathcal{S}$ on the plane. To fix this we divide the overly large torus fundamental domain vertically to obtain two tall thin rectangles. One of these rectangles, such as the shaded region in Fig. 6.10c, is a fundamental domain for the $\mathcal{S}$ action. Some experimentation with orbits reveals the gluings for the edges of this rectangle, also indicated in Fig. 6.10c. By folding the shaded rectangle over the

[5] A group of symmetries, in fact.

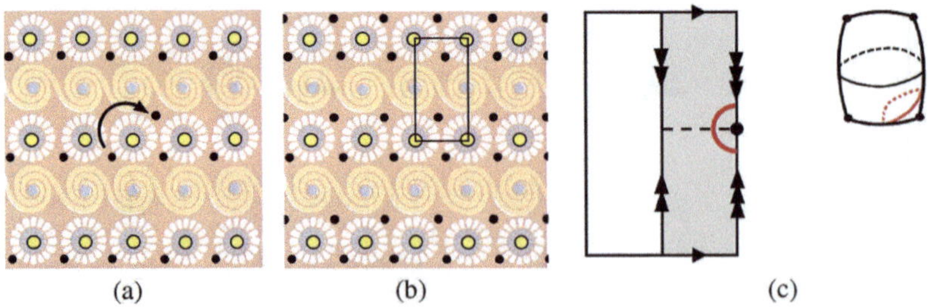

Fig. 6.10 Obtaining the pillow orbifold from translations and rotations. (a) 180° rotational symmetry. (b) Orbits with rotations added. (c) Fundamental domain for a pillow

horizontal dotted line and gluing the edges as indicated, we obtain the quotient of the plane by the symmetries in $\mathcal{S}$. In this case the quotient is shaped like the surface of a pillow, as drawn to the right in Fig. 6.10c.

Note that the pillow surface has four conical corners at which the surface fails to bend smoothly. The presence of these *singular points* – singular in the sense that at these points the surface fails to be smooth – is what makes the pillow surface an *orbifold*[6] rather than a manifold. But how did these singular points arise? They come from points in the fundamental domain that fail to be surrounded by a full 360° angle. The red semicircle drawn within the shaded rectangle in Fig. 6.10c indicates that the black point at its center only has a 180° angle around it in the fundamental domain. Upon folding and gluing the fundamental domain, this point is surrounded by a cone with cone angle 180°. This is indicated by the red circle drawn around the lower right hand corner of the pillow surface in Fig. 6.10c. Note that all points in the fundamental domain of the torus are surrounded by a full 360° angle. For example, the yellow points at the corners of the torus fundamental domain in Fig. 6.8c are all glued together in the quotient. Each of them contributes 90° to the angle around that single (glued) point, yielding a smooth 360° angle in the quotient surface. The torus has no singular points.

One final detail to notice about the examples above is that a 180° rotation, such as the one indicated in Fig. 6.10a, leaves the central point of the rotation in its original position. However, translations leave no points unmoved. Singular points in an orbifold arise in exactly this way, that is, when a symmetry has a fixed point. The full set of symmetries that fixes a particular point is called the *isotropy group* of the corresponding point in the orbifold. In the example of the pillow orbifold the central yellow point is fixed by both the identity symmetry (which leaves the plane in place) and the 180° rotation. Those familiar with group theory will see that the central yellow point, and its corresponding point in the pillow orbifold, has isotropy group the cyclic group of order two: $\mathbb{Z}_2$.

[6] The word orbifold reflects the fact that we *fold* up a fundamental domain according to the *orbits* that define it.

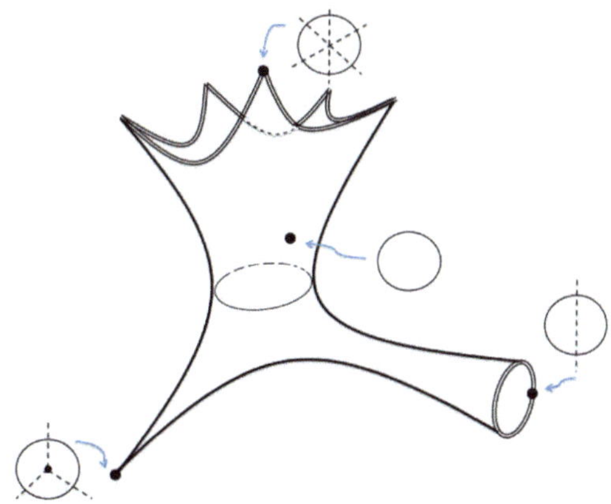

Fig. 6.11 Two-dimensional orbifold with a variety of singular points

In general an orbifold is exactly the same as a manifold, except that an orbifold may have singular points that arise as above – which is by quotients under symmetries where there is not quite enough angle to form a smooth point. Figure 6.11 shows a more general two-dimensional orbifold. Edges drawn with a double line contain singular points called *mirror points*. They correspond to the folded edge of a quotient of the plane by a reflection across a line and also have isotropy group $\mathbb{Z}_2$. The black dot in the lower left of the figure is a *cone point* such as the ones we saw on the corners of the pillow orbifold. This cone is very narrow which suggests that its isotropy group is a higher-order cyclic group $\mathbb{Z}_n$, $n > 2$, generated by a rotation through a smaller angle $\frac{2\pi}{n}$. The black dot at the top of the orbifold is in a neighborhood modeled off of the quotient of a regular polygon by its full set of symmetries. Because the set of symmetries of a regular polygon is called a dihedral group, this type of singular point is called a *dihedral point*. The surface bends smoothly at the black point in the center of the image, and so this point is *nonsingular*. Most points in an orbifold will be nonsingular. Nonsingular, cone, dihedral, and mirror points are the only types of points that appear in a two-dimensional orbifold. Because more symmetries are available in higher dimensions, higher-dimensional orbifolds can have a wide variety of singular points.[7]

6.4.2 *Can You Hear the Shape of an Orbifold?*

The fact that the singular points of an orbifold arise from groups of symmetries allows the rules of calculus to transfer to orbifolds in nearly the same way they carry over to manifolds. The Laplace and Dirichlet-to-Neumann operators can be defined for orbifolds, and the spectra of these operators behave as they do for manifolds:

[7] Many more examples of symmetries and orbifolds can be found in the visually rich text [9].

Each is a sequence of nonnegative numbers that grows to infinity with only finitely many repetitions. Thus, all of the questions from the spectral geometry of manifolds can also be asked for orbifolds.

Of particular interest in the orbifold setting is what aspects of the singular set of an orbifold can be determined from spectral information. One fundamental question is whether or not you can hear the presence of singular points. In other words, can a smooth manifold have the same Laplace spectrum as an orbifold with singular points? This question has not yet been answered. To get a better feeling of what the Laplace spectrum does and does not tell us about the singular points in an orbifold, let us pause to take a careful look at an example of Laplace isospectral orbifolds. This example comes from my joint paper with Shams et al. [23]. This section assumes the reader has had a course in linear algebra and abstract algebra.

A major ingredient in the example is Sunada's theorem, a powerful tool for constructing Laplace isospectral manifolds (or orbifolds). This is the theorem used to construct the isospectral planar domains in Fig. 6.4, for example. Before stating Sunada's theorem we need the following definition.

Definition 6.1 Subgroups H_1 and H_2 of a finite group G are *almost conjugate* if for every $h \in H_1$ there is an element $g \in G$ so that $ghg^{-1} \in H_2$.

Theorem 6.1 (Sunada's Theorem[8]) *Let H_1 and H_2 be almost conjugate subgroups of a finite group G. Let M be a manifold on which G acts by symmetries. Then the quotient space of M by the symmetries in H_1 has the same Laplace spectrum as the quotient space of M by the symmetries in H_2.*

Note that if one were to use conjugate subgroups (so H_1 and H_2 for which $gH_1g^{-1} = H_2$ for some $g \in G$) instead of almost conjugate subgroups, the quotient spaces would be exactly the same. Thus, we would fail to get an interesting example of two *different* spaces with the same Laplace spectrum. With almost conjugate subgroups, it is possible to get different quotient spaces; however, one must do the additional step of proving they are different.

Take G to be the symmetric group of permutations on 27 letters. Let $H_1 = \mathbb{Z}_3 \times \mathbb{Z}_3 \times \mathbb{Z}_3$ and let H_2 be the mod-3 Heisenberg group defined as the set

$$H_2 = \left\{ \begin{bmatrix} 1 & c & a \\ 0 & 1 & b \\ 0 & 0 & 1 \end{bmatrix} : a, b, c \in \mathbb{Z}_3 \right\}$$

together with the operation of matrix multiplication. Cayley's theorem states that every finite group of order n is isomorphic to a subgroup of the symmetric group of permutations on n letters. Because both H_1 and H_2 have order 27, we can view H_1 and H_2 as subgroups of G. Some work with cycle decompositions of

[8] For those with more background, the precise statement is: Let H_1 and H_2 be almost conjugate subgroups of a finite group G. Let M be a compact Riemannian manifold on which G acts faithfully on the left by isometries. Then the quotient spaces $H_1\backslash M$ and $H_2\backslash M$ are isospectral as orbifolds.

permutations (see [2] for more) shows that H_1 and H_2 are almost conjugate. They are not conjugate, for if they were, they would be isomorphic; however, H_1 is abelian and H_2 is not.

We have three groups that we can use in an application of Sunada's theorem. Now let us find a manifold M upon which the permutation group G acts by symmetries. The n-dimensional unit sphere, denoted S^n, is the set of all points in $\mathbb{R}^{n+1}$ that are distance one from the origin. That is,

$$S^n = \{(x_1, x_2, \ldots, x_{n+1}) \in \mathbb{R}^{n+1} : x_1^2 + x_2^2 + \cdots + x_{n+1}^2 = 1\}.$$

So the one-dimensional sphere S^1 is the familiar unit circle as it consists of the points (x_1, x_2) in the plane for which $x_1^2 + x_2^2 = 1$. Similarly, the two-dimensional unit sphere S^2 consists of all points (x_1, x_2, x_3) in $\mathbb{R}^3$ for which $x_1^2 + x_2^2 + x_3^2 = 1$. This is the surface of a ball of radius one in three-dimensional space. (Note that the dimension of the sphere refers to its dimension as a manifold, not to the dimension of the ambient space in which it sits.) The manifold M that we will use in Sunada's theorem is the 26-dimensional unit sphere S^{26}.

To see how G acts on S^{26} by symmetries, let us begin with the more familiar setting of S^2 sitting inside $\mathbb{R}^3$. Imagine you are holding a basketball such as the one in Fig. 6.12. One of the seams on the basketball is a great circle that wraps vertically around the ball, through its north and south poles. Reflection of the ball across that seam is a symmetry of the sphere that preserves the seam pattern of the basketball. This symmetry can be viewed as a linear transformation on $\mathbb{R}^3$. Specifically, first assume the coordinate axes are positioned in the standard way, as indicated in Fig. 6.12. Observe where the reflection sends the standard basis vectors: $\mathbf{e}_1$ and $\mathbf{e}_2$ switch places under this reflection and $\mathbf{e}_3$ is not moved. Using this, we see that the linear transformation $T : \mathbb{R}^3 \to \mathbb{R}^3$ corresponding to this reflection can be represented using matrix multiplication $T(\mathbf{v}) = A\mathbf{v}$ where

$$A = \begin{bmatrix} 0 & 1 & 0 \\ 1 & 0 & 0 \\ 0 & 0 & 1 \end{bmatrix}.$$

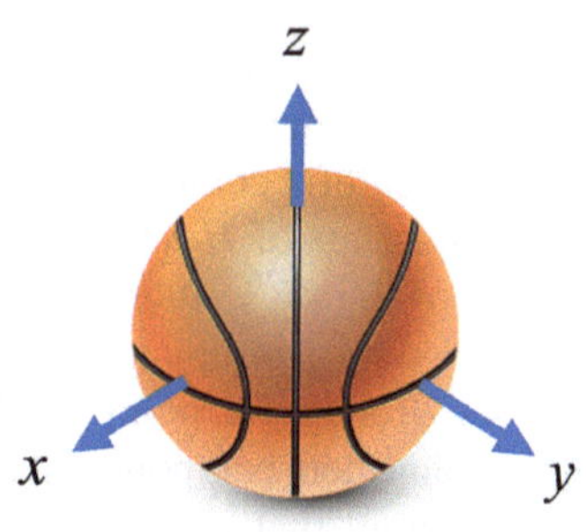

Fig. 6.12 A familiar two-dimensional sphere [20]

The column vectors of A are the standard basis vectors for $\mathbb{R}^3$, but instead of being written in the usual order $\mathbf{e}_1, \mathbf{e}_2, \mathbf{e}_3$, they are written with the first two vectors switched $\mathbf{e}_2, \mathbf{e}_1, \mathbf{e}_3$. Taking the permutation $\sigma = (12)$ in the symmetric group of permutations on three letters, another way to write the column vectors of A is $\mathbf{e}_{\sigma(1)}, \mathbf{e}_{\sigma(2)}, \mathbf{e}_{\sigma(3)}$. We have taken a symmetry of the sphere and characterized it as a matrix whose columns are a permutation of the standard basis vectors of $\mathbb{R}^3$. Any permutation in the symmetric group on three letters corresponds to a symmetry of the sphere in this way. For example, the permutation $\omega = (123)$ corresponds to a rotation of the sphere around the line in $\mathbb{R}^3$ that contains the origin and the point $(1, 1, 1)$. Thus, we have defined an action of the group of permutations on three letters on $\mathbb{R}^3$ for which each permutation acts on the sphere S^2 by a symmetry.

To finish the setup needed to apply Sunada's theorem, we scale this action on $\mathbb{R}^3$ up to $\mathbb{R}^{27}$. That is, take a permutation $\sigma \in G$. So σ is a permutation of the numbers $1, 2, 3, \ldots, 27$. Define matrix A as the matrix that has column vectors $\mathbf{e}_{\sigma(1)}, \mathbf{e}_{\sigma(2)}, \ldots \mathbf{e}_{\sigma(27)}$. Then the linear transformation on $\mathbb{R}^{27}$ given by $T(\mathbf{v}) = A\mathbf{v}$ restricts to a symmetry of the sphere S^{26}. In this way our big group G acts on S^{26} by symmetries, as do the almost conjugate subgroups H_1 and H_2. Applying Sunada's theorem we conclude that the quotient of S^{26} by the symmetries in H_1 and the quotient of S^{26} by the symmetries in H_2 have the same Laplace spectra. We will denote these quotient spaces by $H_1 \backslash S^{26}$ and $H_2 \backslash S^{26}$, respectively.

These Laplace isospectral quotient spaces are exciting! Recall the question that led us to them: What does the Laplace spectrum tell us, or fail to tell us, about the singular set of an orbifold? One answer to this question is immediately available. Consider the points

$$\pm \frac{1}{\sqrt{27}}(1, 1, \ldots, 1)$$

on the sphere S^{26}. A permutation of the basis vectors of $\mathbb{R}^{27}$ does not affect these points because all of the coordinates of the points are identical. So every symmetry in H_1 and every symmetry in H_2 fix both of these points. This implies $H_1 \backslash S^{26}$ has exactly two points with isotropy group H_1. No point in $H_2 \backslash S^{26}$ can have isotropy group H_1 because all of the symmetries in H_1 simply are not present in H_2. Similarly, there are two points in $H_2 \backslash S^{26}$ that have isotropy group H_2, and no such points exist in $H_1 \backslash S^{26}$. What this means is that the Laplace spectrum of an orbifold does not determine the type of isotropy group that may appear in the orbifold. In other words, *one cannot hear the isotropy groups present in an orbifold.*

Some careful computation with these examples also shows that the way the set of singular points of $H_1 \backslash S^{26}$ sits inside $H_1 \backslash S^{26}$ is different from the way the singular points of $H_2 \backslash S^{26}$ sit inside $H_2 \backslash S^{26}$. For example, the set of singular points in $H_1 \backslash S^{26}$ with isotropy group $\mathbb{Z}_3 \times \mathbb{Z}_3$ has thirteen connected components, while the set of singular points with the same isotropy group in $H_2 \backslash S^{26}$ has at most four connected components. Here "connected component" is a term from point-set topology that refers to a set of points in a topological space that cannot be separated into two nonoverlapping open sets. For example, the set of mirror points in the orbifold in Fig. 6.11 is made up of six connected components. Five of these components are in the "crown" shape at the top of the orbifold and one is the "mirror circle" at the end of the orbifold on the bottom right corner of the figure.

As mathematicians, once we get an interesting result we look for ways to extend it in new directions. Is it possible to construct more than two orbifold quotients of the sphere that are Laplace isospectral and have points with distinct isotropy? How many more: ten? a hundred? infinitely many? To answer this question, Shams, Webb, and I constructed arbitrarily large finite families of sphere quotients with these properties. However, a result from my dissertation work [25] shows that this is as bad as it can get. Using the fact that the unit sphere has constant curvature one, it is possible to show that any collection of isospectral orbifold quotients of the sphere can have points with only *finitely* many different types of isotropy group. So we cannot have an infinite list of isospectral orbifold quotients of the sphere where each orbifold on the list has a point with distinct isotropy group; there are not enough isotropy groups to choose from.

Paper [25] was the first publication to derive information about the singular points of an orbifold from the orbifold's Laplace spectrum. Since then many other interesting questions have been posed and some have been resolved. In the example above you might have noticed that the groups H_1 and H_2 have the same number of elements. Thus, the maximal isotropy groups (those of largest order) that appear in $H_1\backslash S^{26}$ and $H_2\backslash S^{26}$ have the same number of elements. Given this it is natural to ask, "Can we hear the order of the largest isotropy group of an orbifold?" This question was resolved in the negative by Rossetti et al. [22].

6.4.3 *The Steklov Spectrum of an Orbifold*

My 2015 Women in Geometry Workshop team[9] established the Steklov eigenvalue problem for orbifolds. This entails defining the Dirichlet-to-Neumann operator on an orbifold with boundary and confirming that despite the singular points that might be present in an orbifold, the operator is still nice enough to have an eigenvalue spectrum. In working with the Steklov spectrum of an orbifold, a first example to consider is the folded disk. A folded disk is obtained by taking a disk of radius R and acting on it by the symmetry of reflection across a diameter. The quotient of the disk by this reflection is a half disk – a semi-circle with a straight edge of mirror points and a curved edge of orbifold boundary points. Four half disks are shown in Fig. 6.13. A waveform on the half disk unfolds into a waveform on the full disk that is not changed by the reflection symmetry. Figure 6.6 shows an example. If you cut the waveform shown there by the xz-plane, you can see that reflection across the plane brings the waveform back to itself. It turns out that only half of the waveforms

[9] The full team is myself, Teresa Arias-Marco, Emily Dryden, Carolyn Gordon, Asma Hassannezhad, and Allie Ray.

on the disk possess this reflective symmetry. Because of this one can show that the Steklov spectrum of the folded disk of radius R is $0, \frac{1}{R}, \frac{2}{R}, \frac{3}{R}, \ldots$.

Figure 6.13 shows an example of Steklov isospectral two-dimensional orbifolds that have different shapes. Perhaps this sounds strange as Fig. 6.13 shows six distinct objects, not two. For reasons that will become clearer soon, we take the first three of the objects in Fig. 6.13 to be a single *disconnected* orbifold and the last three objects in Fig. 6.13 to be the other orbifold which is also *disconnected*. To be very specific, the first of the two orbifolds is the union of two half disks of radius R and a circle of radius r, with $r < R$. This first orbifold is on the left side of Fig. 6.13. The second orbifold is the union of one circle of radius R and two half disks of radius r, again with $r < R$, shown on the right side of the figure. The Steklov spectrum of an orbifold with more than one component is the union of the Steklov spectra of each component. So the Steklov spectrum of the orbifold on the left is

$$\left\{0, \frac{1}{R}, \frac{2}{R}, \frac{3}{R}, \ldots\right\} \cup \left\{0, \frac{1}{R}, \frac{2}{R}, \frac{3}{R}, \ldots\right\} \cup \left\{0, \frac{1}{r}, \frac{1}{r}, \frac{2}{r}, \frac{2}{r}, \frac{3}{r}, \frac{3}{r}, \ldots\right\}.$$

The Steklov spectrum of the orbifold on the right is

$$\left\{0, \frac{1}{R}, \frac{1}{R}, \frac{2}{R}, \frac{2}{R}, \frac{3}{R}, \frac{3}{R}, \ldots\right\} \cup \left\{0, \frac{1}{r}, \frac{2}{r}, \frac{3}{r}, \ldots\right\} \cup \left\{0, \frac{1}{r}, \frac{2}{r}, \frac{3}{r}, \ldots\right\}.$$

Comparing these sets shows that these two orbifolds are Steklov isospectral.

Generalizing techniques from [11], my Women in Geometry team showed that the Steklov spectrum of a two-dimensional orbifold with boundary determines the boundary of the orbifold up to an equivalence relation. We can detect the length of boundary components, as well as whether or not the boundary components are circular or semicircular, except for orbifolds whose boundary includes components the match the form of the boundary of the orbifolds in Fig. 6.13. In the latter case, we cannot use the Steklov spectrum to tell if the boundary looks like the boundary of the orbifold on the left of the figure, or if it looks like the boundary of the orbifold on the right of the figure.

Research into the Steklov spectral geometry of orbifolds is still relatively new and I look forward to its further development.

Fig. 6.13 Steklov isospectral orbifolds

6.5 Other Research Directions

My work as a college educator has inspired two other directions in my research program. One is in spectral graph theory and the other is in math biology education.

6.5.1 *Spectral Graph Theory*

In graph theory, the word graph refers to a network of vertices connected by lines called edges. Figure 6.14 gives an example of a graph with four vertices and six edges. A basic tool in graph theory is the adjacency matrix of a graph. A one in the jth column of the ith row of the adjacency matrix means that there is an edge in the graph connecting vertex v_i to vertex v_j. A zero in the same location of the adjacency matrix indicates that there is no edge between vertex v_i and vertex v_j. For example, the adjacency matrix of the graph in Fig. 6.14 is

$$\begin{bmatrix} 0 & 1 & 1 & 1 \\ 1 & 0 & 1 & 1 \\ 1 & 1 & 0 & 1 \\ 1 & 1 & 1 & 0 \end{bmatrix}.$$

In a directed graph (see Fig. 6.15, for example), each edge is marked with an arrow indicating the direction of the edge. The adjacency matrix of a directed graph has a one in the ijth entry only when an edge goes from the ith vertex to the jth vertex; the entry is zero otherwise.

Just as manifolds and orbifolds are studied using the eigenvalues of associated operators, graphs are studied using the eigenvalues of matrices associated to the graph. One approach is to take the graph spectrum to be the list of eigenvalues (with repetitions allowed) of the adjacency matrix of the graph. There are interesting connections between the study of the spectral properties of a manifold and the study of the spectral properties of a graph. For example, one can ask if the Laplace spectrum of a manifold determines the volume of the manifold. The analogous question in spectral graph theory is to ask if the spectrum of a graph determines the number of vertices in the graph. (The answer to both questions is yes.) The

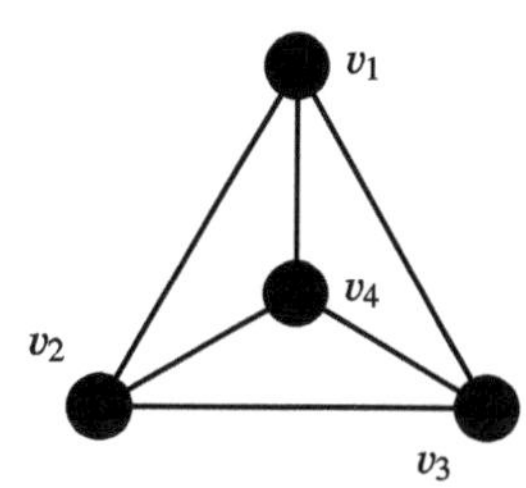

Fig. 6.14 A graph with four vertices and six edges

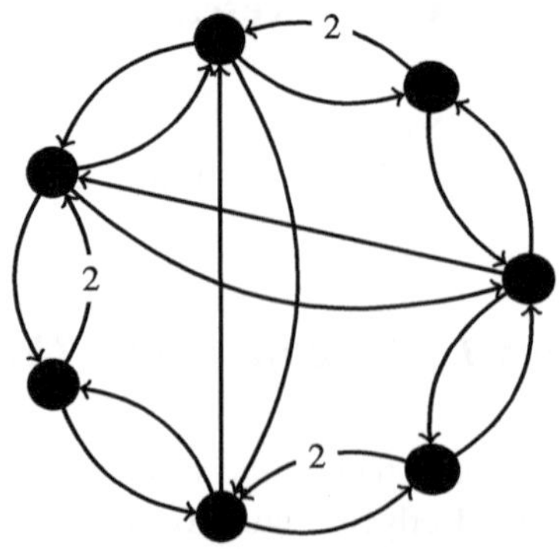

Fig. 6.15 An orbigraph with seven vertices [10]

strong relationship between spectral geometry and spectral graph theory has been a subject of great interest (see [3, 4, 27], for example).

Because my contributions to spectral geometry focus on the properties of orbifolds, my motivating question in spectral graph theory is: What is the graph theoretic analogue of an orbifold and what are its spectral properties? In [10] undergraduate collaborators[10] and I propose a definition of an orbifold graph that mirrors the definition of an orbifold. The directed graph in Fig. 6.15 is an example of an orbifold graph. Note that edges in this directed graph labeled with the number 2 count as two directed edges when writing down the adjacency matrix of the graph. Key features of an orbifold graph are that the number of edges leaving a vertex is always the same (that number is 3 in Fig. 6.15) and that if an edge goes from v_i to v_j, there must be a "return" edge that goes from v_j to v_i. A vertex in an orbifold graph is called *singular* if an edge leaving that vertex is labeled with a number.

One result that we obtain is that the number of singular points in an orbifold graph is bounded above and below by constants determined only by the spectrum of the orbifold graph's adjacency matrix. Thus, we can use the spectrum of the orbifold graph to understand the singular set of the graph, in analogy to spectral results about orbifolds in the setting of the Laplace or Dirichlet-to-Neumann operator.

6.5.2 Discipline-Based Education Research

The rise of scientific computing has opened up a new world of possibilities to those who know both mathematics and biology. This connection is so strong that mathematics has even been described as biology's "new microscope" [5]. The increasing emphasis on mathematics in biology has put pressure on undergraduate biology educators to incorporate more mathematical methods into the biology curriculum. In 2013 a team including myself, undergraduate chemistry and biology faculty,

[10] Kathleen Daly, Colin Gavin, Gabriel Montes de Oca, Diana Ochoa, and Sam Stewart.

educational psychologists, and a college institutional research staff member[11] formed to better understand how to assess the progress undergraduate institutions are making in bringing more math into biology. My team developed a multiple choice assessment called the *BioSQuaRE*. This assessment measures student proficiency with sophomore-level quantitative biology. Our paper [26] describes the development of the assessment and serves as a roadmap for other discipline-based education researchers interested in developing robust assessments of courses or programs.

6.6 Future Work

In her 2019 Glamour Women of the Year award speech, soccer star Megan Rapinoe reflected on how her career was built on hard work and talent, as well as the work of many others in building a culture where a queer woman athlete can be seen internationally as a symbol of power and success. She urged members of the awards ceremony audience to "throw your ladders down" to continue the process of extending opportunities to all. Although being a tenured mathematician at a US college is fabulous in a more modest way, my vision for my future work centers on throwing down ladders. From the leadership of the women of *Hidden Figures* to the storytelling of author Leslie Feinberg, my position in this time in space is a privilege that stands on decades of work by those who came before. As I pursue future directions of research, teaching, and service, I look forward to throwing down ladders. I hope that my work with the *BioSQuaRE* will eventually help the profession understand how to decrease the barrier quantitative skills pose to students seeking life science degrees. Work in spectral geometry with undergraduates and my Women in Geometry teams will advance mathematical knowledge while providing opportunities to mentor. From this mid-career vantage point, I look forward to seeing how these efforts develop.

References

1. Boyce, W.E., DiPrima, R.C.: Elementary Differential Equations and Boundary Value Problems. John Wiley & Sons, New York-London-Sydney (1965)
2. Brooks, R.: Constructing isospectral manifolds. Am. Math. Monthly **95**(9), 823–839 (1988)
3. Brooks, R.: Non-Sunada graphs. Ann. Inst. Fourier (Grenoble) **49**(2), 707–725 (1999)
4. Chung, F.R.K.: Spectral graph theory. In: CBMS Regional Conference Series in Mathematics, vol. 92. Published for the Conference Board of the Mathematical Sciences, Washington, DC; by the American Mathematical Society, Providence (1997)

[11] The full team is myself, Laura Ziegler, Tabassum Haque, Laura Le, Marcelo Vinces, Gregory K. Davis, Andy Zieffler, Peter Brodfuehrer, Marion Preest, Jason Belisky, Charles Umbanhowar, Jr., and Paul J. Overvoorde.

5. Cohen, J.: Mathematics is biology's next microscope, only better; biology is mathematics' next physics, only better. PLOS Biol. **2**(12), e439 (2004)
6. Colbois, B., El Soufi, A., Girouard, A.: Isoperimetric control of the Steklov spectrum. J. Funct. Anal. **261**(5), 1384–1399 (2011)
7. Colbois, B., Girouard, A., Gordon, C., Sher, D.: Some recent developments on the Steklov eigenvalue problem. Rev. Mat. Complut. **37**(1), 1–161 (2024)
8. Commons Wikimedia: Public Domain (2007). https://commons.wikimedia.org/wiki/File:Isospectral_drums.png. Accessed 17 Sep 2024
9. Conway, J.H., Burgiel, H., Goodman-Strauss, C.: The Symmetries of Things. A K Peters, Wellesley (2008)
10. Daly, K., Gavin, C., Montes de Oca, G., Ochoa, D., Stanhope, E., Stewart, S.: Orbigraphs: a graph-theoretic analog to Riemannian orbifolds. Involve **12**(5), 721–736 (2019)
11. Girouard, A., Parnovski, L., Polterovich, I., Sher, D.: The Steklov spectrum of surfaces: asymptotics and invariants. Math. Proc. Camb. Philos. Soc. **157**(3), 379–389 (2014)
12. Girouard, A., Polterovich, I.: Spectral geometry of the Steklov problem (survey article). J. Spectr. Theory **7**(2), 321–359 (2017)
13. Gordon, C., Webb, D.: You can't hear the shape of a drum. American Scientist **84**, 46–55 (1996)
14. Gordon, C., Webb, D., Wolpert, S.: Isospectral plane domains and surfaces via Riemannian orbifolds. Invent. Math. **110**(1), 1–22 (1992)
15. Harthong, J., Rubenthaler, H.: Contre exemple de Milnor, p. 30 (1973)
16. Janssen, P., Lockyer, N.: C. R. Acad. Sci. Paris **67**, 836–841 (1868)
17. Jones, O.: The grammar of ornament. Quaritch (1910). http://digital.library.wisc.edu/1711.dl/DLDecArts.GramOrnJones
18. Karhunen, K., Seppänen, A., Kaipio, J.P.: Adaptive meshing approach to identification of cracks with electrical impedance tomography. Inverse Probl. Imaging **8** (1), 127–148 (2014)
19. Kuznetsov, N., Kulczycki, T., Kwaśnicki, M., Nazarov, A., Poborchi, S., Polterovich, I., Siudeja, B.o.: The legacy of Vladimir Andreevich Steklov. Notices Amer. Math. Soc. **61**(1), 9–22 (2014)
20. Mazenl77 - FindIcons: CC BY 3.0. https://commons.wikimedia.org/w/index.php?curid=15721819
21. Milnor, J.: Eigenvalues of the Laplace operator on certain manifolds. Proc. Nat. Acad. Sci. U.S.A. **51**, 542 (1964)
22. Rossetti, J.P., Schueth, D., Weilandt, M.: Isospectral orbifolds with different maximal isotropy orders. Ann. Global Anal. Geom. **34**(4), 351–366 (2008)
23. Shams, N., Stanhope, E., Webb, D.L.: One cannot hear orbifold isotropy type. Arch. Math. (Basel) **87**(4), 375–384 (2006)
24. Karhunen, K., Seppänen, A., Lehikoinen, A., Monteiro, P.J.M., and Kaipio, J.P. : Crack identification in concrete with Electrical Resistance Tomography. In: Proceedings of 5th European Congress on Computational Methods in Applied Sciences and Engineering (ECCOMAS 2008), Venice (2008)
25. Stanhope, E.: Spectral bounds on orbifold isotropy. Ann. Global Anal. Geom. **27**(4), 355–375 (2005)
26. Stanhope, E., Ziegler, L., Haque, T., Le, L., Vinces, M., Davis, G., Zieffler, A., Brodfuehrer, P., Preest, M., Belisky, J., Umbanhowar Jr., C., Overvoorde, P.: Development of a biological science quantitative reasoning exam (BioSQuaRE). CBE Life Sci. Educ. **16**(4) (Winter 2017)
27. Sunada, T.: Riemannian coverings and isospectral manifolds. Ann. Math. (2) **121**(1), 169–186 (1985)
28. Kokich, R.: Solar spectrum Balmer series uncalibrated accessed: August 5 2024 (2016). https://www.flickr.com/photos/140097441@N02/26070943080. Marked with Public Domain Mark 1.0

Chapter 7
Continued Fractions Connecting Number Theory, Dynamical Systems, and Hyperbolic Geometry

Claire Merriman

Abstract Continued fractions are a way to represent numbers as $a_0 + \cfrac{e_1}{a_1 + \cfrac{e_2}{\ddots}}$.

They appear in various areas of mathematics, from number theory to dynamical systems to geometry. I will describe functions which generate these fractions. The continued fraction expansions provide a nice description of paths on the geometric surfaces. The geometric properties of these pictures help us to describe patterns in the continued fraction expansions, and the continued fraction expansions provide a compact description of the geometry.

Various types of continued fractions correspond to different rules for the numerators and denominators. The first example is the "regular" or "simple" continued fractions, where all of the numerators, e_i, are 1. I then show how the dynamics and geometry change when all of the numerators are ± 1 and the denominators, a_i, are all even.

Mathematical Background: Since I am using functions to generate continued fractions, you should be comfortable with function composition, as well as preimages of points. I will refer to functions that are onto (every point has a preimage) and one-to-one (different points have different images) as invertible. There is also a lot of adding and multiplying fractions and use of the fact that $\frac{1}{\frac{1}{a}} = a$.

For the geometry, you should be comfortable with rotating objects in space. Complex numbers will also show up throughout. If you are unfamiliar with complex variables, you can think of the point $z = x + iy$ as (x, y) in the plane. Circles of radius r centered at $x + iy$ in the complex plane can be written as $x + iy + re^{it}$ where $t \in (0, 2\pi)$. Here, all of our circles are centered on the x-axis, so $y = 0$.

Keywords Continued fractions · Hyperbolic plane · Tessellation · Cutting sequences · Number theory · Geometry · Numerical approximation · Insertion

C. Merriman (✉)
George Mason University, Fairfax, VA, USA
e-mail: emerrima@gmu.edu

A. Brisbin et al. (eds.), *Research Connections*, Association for Women in Mathematics Series 36, https://doi.org/10.1007/978-3-031-81217-0_7

and singularization · Even continued fractions · Regular continued fractions · Modular surface · Invariant measure · Dynamical system

7.1 Biography

7.1.1 How I Got Here

When I was 16, I decided to attend Johns Hopkin's Center for Talented Youth (CTY). I knew I was good at math, so I signed up for number theory. Suddenly, a world of formulas was transformed into proofs. Figuring out where mathematics comes from and why it works was way more exciting than manipulating algorithms. That early introduction to proofs inspired me to go into math. It was also my first introduction to math summer programs, which would go on to tremendously influence my graduate school path.

The Carleton College Summer Math Program for Women (SMP) was the other summer program that influenced my graduate school path. I started college at a tiny liberal arts school. Because I had the opportunity to double up my math classes in high school, I was immediately in more advanced classes with fewer than 10 people. One professor suggested I apply for SMP. The courses looked interesting, and being paid to study math was exciting, so I applied. Because I had not had a great experience at an all-girls school, I was not initially thrilled that the program was for women. I am grateful that SMP showed me the value of an all-women program. Mentoring is one of the most important ways that students are supported through school, from kindergarten through graduate school. Women are less likely to be mentored naturally, and programs like SMP step in to fill that gap. I am grateful for all of the support and advice the SMP community has given me through grad school, especially for supporting my way of being a mathematician.

Between my experiences at CTY and SMP, I knew the value of math programs outside of the standard curriculum. There were many reasons I chose the University of Illinois for graduate school. I was worried that I would struggle to find an advisor I worked well with in a department with fewer professors. I also knew the U of I ran many outreach activities in the local community. My first semester of grad school I began helping with our Association of Women in Mathematics (AWM) outreach programs. Both semesters, we hosted a series of two-hour activities for middle school girls and an all-day program for high school girls.

That winter at the Joint Math Meetings, I was talking to a group of Illinois grad students about our department's outreach efforts, and we realized we did not run any programs in the summer. So Melinda Lanius, another SMP alum, and I decided to change that. We won a $1500 grant from the university, convinced the Department of Mathematics to pay for lunch, wrote lesson plans, and found eight other grad students to help as classroom assistants. I even convinced a professor to give me a one-month assistantship for the work. Through some whirlwind marketing, we managed to receive 60 applications from local high school students interested in doing math for a week in the summer. And then, the first week of August 2015, we ran the first Summer Illinois Math (SIM) Camp. In the seven months between the joint meetings and camp, we did not have time to stop and think about the impact of what we were doing. I am pretty sure if we did, I would have been overwhelmed by imposter syndrome. That first summer, I taught a number theory and cryptography course based on my CTY textbook, and Melinda taught low-dimensional topology based on our SMP notes. We modeled our courses on those that had influenced us the most, because we wanted others to have those opportunities. Summer 2018 was the fourth year of SIM Camp and our first summer offering three camps.

SIM Camp remains almost entirely graduate-student-run, with undergraduate assistant instructors. As of 2018, eleven graduate students have been instructors, and four of those are SMP alums.

7.1.2 How I Found My Math

Mathematics PhD programs train students to be mathematicians, but there are many ways to be a mathematician. My way involves getting others excited about math through teaching and outreach, but it also involves research. When I started graduate school, I was not sure what area of mathematics I wanted to study.

After taking Lie theory, an area of algebra, at SMP, I really wanted to learn more about it. I took a Lie theory course my first semester of grad school, and despite spending most of the time feeling very confused, I really enjoyed it and the professor. After seeing him give a talk that I really liked, I asked him to do a reading course, to see if I was interested in working with him. I also applied to be the graduate student mentor on an undergraduate research project that he was leading. At the end of that semester, we sat down to talk about my plans, and I told him that everything that we were working on was too analytic for my tastes. We made a list of what type of math I wanted to do and who I should email about reading courses.

Ultimately, I ended up with a professor who does even more analysis, but I still do not. At least, all of the analysis I do is either handled by the geometry or only uses integrals that are easier than some of our calc 2 assignments. But we have similar approaches to working on math, even if we have different approaches to the actual problem. Sometimes we have two-hour meetings and really work things out, and sometimes we meet for five minutes because we have nothing to say. Anyone who is worried about finding an advisor or thesis problem should talk to a professor or older student they like, either mathematically or personally. Talking to someone with a better feel of the mathematical landscape is incredibly helpful. But it is also important to know it is ok to do something different than your advisor. Having different approaches to problems can be incredibly useful.

My research is between a few areas of mathematics, which I like, because it allows me to work with lots of different people. Sometimes I am a number theorist, sometimes I am a geometer, and frequently I am a dynamicist and ergodic theorist.

Really, I study continued fractions (number theory) by drawing pictures (two-dimensional geometry) and using those pictures to figure out what happens when you apply the same map over and over (dynamics).

7.2 Introduction

7.2.1 What Is a Continued Fraction?

Continued fractions are a way to represent real numbers as

$$x = a_0 + \cfrac{e_1}{a_1 + \cfrac{e_2}{a_2 + \cfrac{e_3}{a_3 + \dots}}},$$

with various restrictions on the e_i and a_i. Typically, the numerators e_i are ± 1 and the denominators a_i are positive integers, but there could be further restrictions. For example, you may have heard that π is approximately $\frac{22}{7}$, which comes from the continued fraction approximation $\pi \approx 3 + \frac{1}{7} = 3.\overline{142857}$. In fact,

$$\pi = 3 + \cfrac{1}{7 + \cfrac{1}{15 + \dots}} \tag{7.1}$$

where $3 + \frac{1}{7+\frac{1}{15}}$ agrees with π to four decimal places. For almost all real numbers, the continued fraction converges faster than the decimal expansion. We can also write π with only even denominators as

$$\pi = 4 - \cfrac{1}{2 - \cfrac{1}{2 - \cfrac{1}{8 + \cfrac{1}{2 - \dots}}}}. \tag{7.2}$$

The rules for different types of continued fraction expansions, such as "every numerator is $+1$" or "every denominator is even," connect to different geometry and matrix groups. This chapter will explain how to connect geometry to the regular continued fractions, where every numerator is $+1$, and the continued fractions with even denominators.

7.2.2 *Preview of Connections with Other Fields of Mathematics*

The continued fraction expansions provide a nice description of paths on the geometric surfaces. As part of the geometric construction, I introduce the hyperbolic plane and describe a tiling made of hyperbolic triangles in Sect. 7.4. The *regular continued fraction expansions* are like the continued fraction expansion of π in Eq. (7.1), where every numerator is $+1$ and every denominator is a positive integer. These continued fractions correspond to something geometers study called a *modular surface*, which is like a sphere where you stretch one point out to infinity and pinch two other points down like the point of a cone. This shape comes from folding the *hyperbolic triangles* in Fig. 7.1 into thirds in a specific way to get a quadrilateral and then gluing adjacent sides together. If you did a similar gluing with a square piece of paper, you would fold along the diagonal to glue the left side to the bottom and the right side to the top and then inflate the resulting shape. You would then have cone points at the 3 remaining corners.

Both the modular surface and the regular continued fractions are related to the group of 2×2 matrices with integer entries and determinant 1. This relation allows us to construct the modular surface, as well as use the regular continued fractions to describe something called *group equivalences*. The *cutting sequence construction for regular continued fractions* in Sect. 7.4.2 let us describe which "straight lines" on the modular surface come back to their starting point, like the equator or longitude lines on a sphere (for example, Series' paper "The modular surface and continued fractions" [14]). Facts about the *invariant measure for the regular continued fractions* also prove that there are lines that hit almost every point on the surface, but never intersect themselves. My goal in this chapter is explaining the geometric construction behind the proofs of these facts, rather than prove them.

The continued fraction expansion of π in Eq. (7.2) is the *even continued fraction expansion*, where every numerator is ± 1 and every denominator is a positive even number. My work connects the even continued fractions to a different modular surface. To get this surface you fold the hyperbolic triangle in half through one of the vertices, which also folds one of the sides in half. Glue the folded side to itself and the other two sides to each other. This gives a sphere with two points dragged

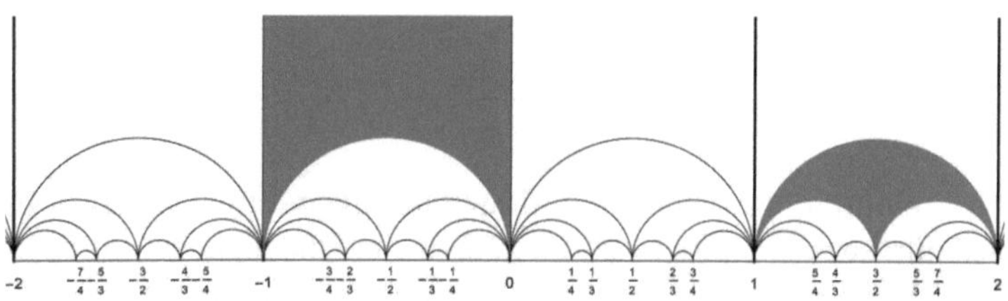

Fig. 7.1 The image used for the geometric description in Sect. 7.4.2. The two hyperbolic triangles with vertices $\{-1, 0, \infty\}$ and vertices $\{1, \frac{3}{2}, 2\}$ are highlighted in grey

to infinity and one cone point. This modular surface is related to determinant 1 matrices of the form $\begin{pmatrix} a & b \\ c & d \end{pmatrix}$, where a, b, c, d are integers and ac and bd are always even. This is a subgroup of the group from the regular continued fractions. Now, the even continued fractions describe group equivalences and straight lines on the surface. Because the invariant measure for the even continued fractions described in Proposition 7.2 can be infinite, it is far more difficult to prove the types of results that tell us there exist lines that visit almost every point without intersecting. However, this fact is already known for modular surfaces related to subgroups of matrices with determinant 1. One of the main ways people prove similar results is by finding a geometric description.

In Sect. 7.3, I describe regular, even, and extended continued fractions as *dynamical systems*. The dynamical systems are what allow us to translate between continued fractions and geometry. I introduce the hyperbolic plane at the beginning of Sect. 7.4 and then define a *tiling* of the hyperbolic plane in Sect. 7.4.1. Finally, I connect the regular continued fraction expansions to the hyperbolic plane using *cutting sequences* in Sect. 7.4.2. I then repeat this process for the even continued fractions in Sect. 7.5. For both the regular and even continued fractions, I describe the continued fraction expansion of two irrational numbers based on how a semicircle connecting these points crosses the tiling from Sect. 7.4.1. In fact, one part of the dynamical system is what we call an *invariant measure*, which I find using the geometric construction. My goal is to explain why you can view continued fractions and paths on these geometric surfaces as essentially the same thing. That is, I can give you continued fractions and you give me the cutting sequence, or I can give you the cutting sequence and you give me the continued fraction.

7.2.3 Computing Continued Fractions

To get the *regular continued fraction expansion* of a number $x > 0$, we start with the largest integer less than x and call this integer a_0. Now we can write $x = a_0 + \frac{1}{r_1}$ for some $r_1 > 1$. To get the next digit, repeat this process for r_1, so a_1 is the largest integer less than r_1 and $x = a_0 + \frac{1}{a_1 + \frac{1}{r_2}}$. If we keep going, we get

$$x = a_0 + \cfrac{1}{a_1 + \cfrac{1}{a_2 + \dots}} = [a_0; a_1, a_2, \dots].$$

It will be helpful to use a running example throughout the chapter. I will use α, an irrational number very close to $\frac{36}{13}$. First, let us find the regular continued fraction expansion of $\frac{36}{13}$. The first step of finding the regular continued fraction expansion of $\frac{36}{13}$ is to write it as $2 + \frac{10}{13} = 2 + \frac{1}{13/10}$. To find the next digit, we rewrite $\frac{13}{10}$ as

$1 + \frac{3}{10} = 1 + \frac{1}{10/3}$. Thus, we get $2 + \frac{1}{1+\frac{1}{10/3}}$. Finally, we rewrite $\frac{10}{3}$ as $3 + \frac{1}{3}$. Thus, we find that $\frac{36}{13} = 2 + \frac{1}{1+\frac{1}{3+\frac{1}{3}}}$. To define α, we replace the last 3 with $3 + z$ for $z \in [0, 1)$, so

$$\alpha := 2 + \cfrac{1}{1 + \cfrac{1}{3 + \cfrac{1}{3 + z.}}} \tag{7.3}$$

When we need an example in $[0, 1)$, I will use $1/\alpha$.

To get the regular continued fraction expansion of negative numbers, we find the continued fraction expansion of the absolute value and then write $x = -[a_0; a_1, a_2, \ldots]$. The regular continued fraction expansion terminates exactly when x is rational, like how the decimal expansion of rational numbers terminate or repeat (see, for example, Chapter 3, Lemmas 3.1 and 3.2 of *Ergodic Theory with a view towards Number Theory* [6]). The regular continued fraction expansions of irrational numbers are unique. However, the regular continued fraction expansion of rational numbers is not unique because the final denominator could be $n = n-1+\frac{1}{1}$. We require that the last denominator is greater than 1 to get a unique continued fraction expansion. The ambiguity of continued fraction expansions is similar to how $1 = 0.99\ldots$, but we typically write the finite decimal expansion. Finally, it is useful to notice that

$$\frac{1}{[a_0; a_1, a_2, \ldots]} = \cfrac{1}{a_0 + \cfrac{1}{a_1 + \cfrac{1}{a_2 + \ldots}}} = \frac{1}{a_0 + [0; a_1, a_2, \ldots]} = [0; a_0, a_1, a_2, \ldots] \tag{7.4}$$

$$\frac{1}{[0; a_1, a_2, \ldots]} = \cfrac{1}{\cfrac{1}{a_1 + \cfrac{1}{a_2 + \ldots}}} = a_1 + \frac{1}{a_2 + [0; a_3, \ldots]} = [a_1; a_2, a_3, \ldots]. \tag{7.5}$$

Notice that if $x \in \left(\frac{1}{k+1}, \frac{1}{k}\right]$, then $x = \dfrac{1}{k+z}$ for some $z \in [0, 1)$. So, we can break up the interval $(0, 1)$ based on the first digit of the regular continued fraction expansion. In Sect. 7.3, I define a function, or *map*, that uses this partition of $(0, 1)$ to generate continued fraction expansions, which we need in order to find the continued fraction expansion of irrational numbers.

7.3 Dynamical Systems

Dynamicists study what happens when we apply the same map over and over. In general, a *dynamical system* consists of a set X, a surjective map (function) $T : X \to X$, and a function μ : subsets of $X \to \mathbb{R}_{\geq 0}$ that measures the size of sets.[1] We call μ a measure and require that for disjoint sets, the size of the union of the sets equals the sum of the sizes of the (individual) sets. In this chapter, all of the measures are integrals of functions. For example, the standard measure on $\mathbb{R}$ is the length of an interval. As an integral, this measure is $\mu([a, b]) = \int_a^b dx = b - a$. Since we are interested in how points $x \in X$ move under T, we say T *acts* on X.

7.3.1 Regular Continued Fractions

We can use a dynamical system to generate the regular continued fractions. Here, $X = [0, 1]$, and $T : [0, 1] \to [0, 1]$ (Fig. 7.2[2]) is given by

$$T(x) = \begin{cases} \frac{1}{x} - k & \text{for } x \in \left(\frac{1}{k+1}, \frac{1}{k}\right] \\ 0 & \text{for } x = 0. \end{cases}$$

We call T the *Gauss map for the regular continued fractions*. The first digit after 0 of the continued fraction expansion of x is k. If we look at what T does to the continued fraction expansion of $x = [0; a_1, a_2, \ldots]$, we get

$$T([0; a_1, a_2, \ldots]) = a_1 + \cfrac{1}{a_2 + \ldots} - a_1 = \cfrac{1}{a_2 + \ldots} = [0; a_2, a_3, \ldots].$$

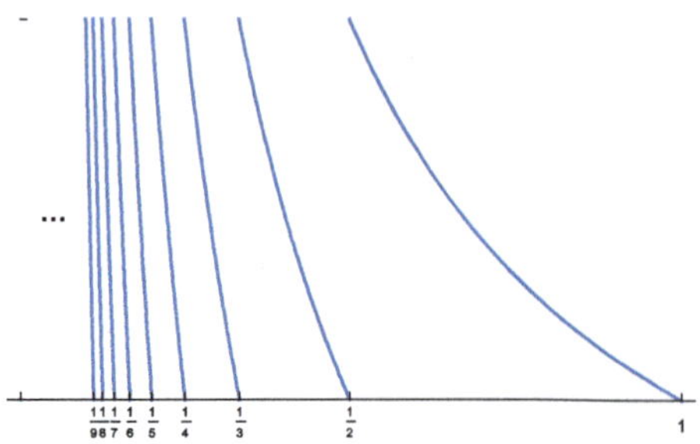

Fig. 7.2 Gauss map T

[1] Dynamical systems also involve a σ-algebra which you can think of as subsets of X.

[2] Figures 7.2, 7.3, and 7.5 appeared in my PhD thesis as Figs. 2.2.1, 2.2.3, and 3.1.1, respectively [11].

So, T deletes the first digit of the continued fraction expansion. In fact, we can write the regular continued fraction expansion of x by repeatedly applying T and recording which integer we subtract.

Using α from Eq. (7.3), we see that $\frac{1}{3} < \frac{1}{\alpha} \approx \frac{13}{36} \leq \frac{1}{2}$. Thus, $T(\frac{1}{\alpha}) = \frac{1}{1/\alpha} - 2$ and the first digit of the continued fraction expansion of $\frac{1}{\alpha}$ is 2. Using the fact that $\alpha - 2 \approx \frac{36}{13} - 2 = \frac{10}{13}$, we see that $\frac{1}{2} < T(\frac{1}{\alpha})) \leq 1$, so $T(T(\frac{1}{\alpha}) = \frac{1}{T(1/\alpha)} - 1$. Thus, the second digit of the continued fraction expansion of $\frac{1}{\alpha}$ is 1. We can continue this process to find all of the digits of $\frac{1}{\alpha}$. Recall from Eq. (7.5) that the continued fraction expansion of $\frac{1}{\alpha}$ also tells us the continued fraction expansion of α. Thus, we can find the continued fraction expansion of all positive real numbers using a dynamical system. For negative numbers, we find the continued fraction expansion of the absolute value.

The last part of the dynamical system is the measure. We want a *T-invariant measure*, which is a function $d\mu = h(x)dx$ so that for any $(a, b) \subseteq [0, 1]$,

$$\int_{T^{-1}(a)}^{T^{-1}(b)} d\mu = \int_a^b d\mu.$$

That is, any set is the same size as its preimage. Since the size of the set does not change, we can more easily study how T moves subsets of $[0, 1]$.

Theorem 7.1 *The probability measure* $\mu([a, b]) = \dfrac{1}{\log 2}\displaystyle\int_a^b \frac{dx}{1+x}$ *is* T*-invariant.*

Proof A full proof is in Chapter 3 of *Ergodic Theory With a View Towards Number Theory* [6], but try proving it for yourself when $a = 0$. First, we find $T^{-1}(0)$ and $T^{-1}(b)$ for a fixed k:

$$0 = \frac{1}{x} - k \quad \text{so} \quad \frac{1}{k} = x,$$

$$b = \frac{1}{x} - k \quad \text{so} \quad \frac{1}{b+k} = x.$$

Thus,

$$T^{-1}([0, b]) = \bigcup_{k=1}^{\infty} \left[\frac{1}{b+k}, \frac{1}{k}\right].$$

We also need that

$$\int_{T^{-1}([0,b])} \frac{dx}{1+x} = \sum_{k=1}^{\infty} \int_{\frac{1}{b+k}}^{\frac{1}{k}} \frac{dx}{1+x}.$$

Now, try to finish the proof yourself. □

There is a no general process for finding the T-invariant measure. Section 7.4.2 will walk through a geometric way to find an invariant measure for the continued fractions. It can be easier for invertible maps, so we are going to define a two-dimensional, invertible version of the Gauss map T, called the *natural extension* $\bar{T}$, on $[0, 1]^2$.

The natural extension $\overline{T}$ will move the first digit of x to the first digit of y. That is, if $(x, y) \in [0, 1]^2$, and

$$(x, y) = \left(\cfrac{1}{a_1 + \cfrac{1}{a_2 + \dots}}, \cfrac{1}{b_1 + \cfrac{1}{b_2 + \dots}} \right), \text{ then}$$

$$\bar{T}(x, y) = \left(\cfrac{1}{a_2 + \cfrac{1}{a_3 + \dots}}, \cfrac{1}{a_1 + \cfrac{1}{b_1 + \dots}} \right).$$

Formally, we define $\bar{T} : [0, 1]^2 \to [0, 1]^2$ by

$$\bar{T}(x, y) = \begin{cases} \left(\frac{1}{x} - k, \frac{1}{k+y}\right) \text{ for } x \in \left(\frac{1}{k+1}, \frac{1}{k}\right] \\ (0, y) \text{ for } x = 0. \end{cases}$$

The first coordinate of $\bar{T}$ is the Gauss map, and the second is the inverse of the Gauss map for a given k.

I will describe the regular continued fractions geometrically in Sect. 7.4.2. First, I am going to describe continued fractions with even denominators. The geometric descriptions of the other types of continued fraction expansions are in Sect. 7.5.

7.3.2 *Even Continued Fractions*

We do not need to require that all numerators are $+1$ to get a valid continued fraction expansion. In this section, we will let our numerators be ± 1, but require that the denominators are all even, which were first introduced by Schweiger [13].

Returning to our example of $\frac{36}{13}$, the first step in finding the even continued fraction expansion is the same, since $2 + \frac{10}{13} = 2 + \frac{1}{13/10}$. However, we can no longer write $\frac{13}{10} = 1 + \frac{1}{10/3}$, since 1 is odd. Instead, we round *up*, getting $\frac{13}{10} = 2 - \frac{7}{10} = 2 - \frac{1}{10/7}$. We repeat this for $\frac{10}{7} = 2 - \frac{1}{7/4}$ and $\frac{7}{4} = 2 - \frac{1}{4}$. Putting this all together, we get

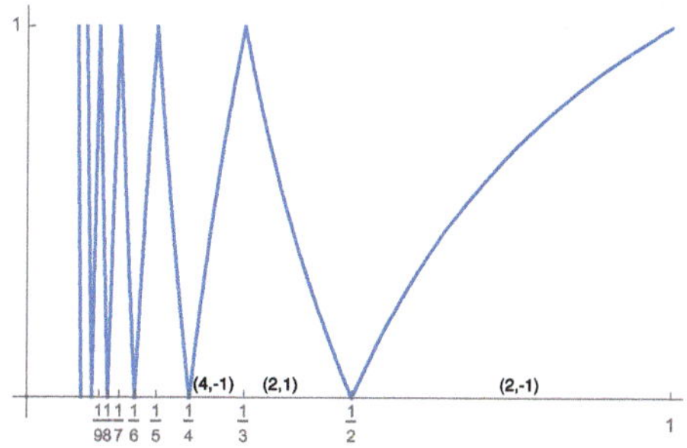

Fig. 7.3 Even Gauss map, with the corresponding digit labeled for the largest three branches

$$\frac{36}{13} = 2 + \cfrac{1}{2 - \cfrac{1}{2 - \cfrac{1}{2 - \cfrac{1}{4}}}} = [\![(2, +1); (2, -1)(2, -1)(2, -1)(2, -1)(4, +1)]\!]. \tag{7.6}$$

The (2, +1) indicates that the 2 was an *underestimate* and we add the remainder, while the (2, −1) indicated the 2 is an *overestimate* and we subtract the remainder. Since we determine this information at the same time, we group the denominator with the *next* numerator. In general, the even continued fraction expansion of real numbers greater than 1 is

$$a_0 + \frac{e_0}{a_1 + \dots} = [\![(a_0, e_0); (a_1, e_1)(a_2, e_2)(a_3, e_3) \dots]\!],$$

where $e_i = \pm 1$, and every a_i is a positive even integer. We write real numbers $x \in (0, 1]$ as

$$x = \cfrac{1}{a_1 + \cfrac{e_1}{a_2 + \cfrac{e_2}{a_3 + \cdots}}} = [\![0; (a_1, e_1)(a_2, e_2)(a_3, e_3) \dots]\!],$$

similar to our notation for the regular continued fractions.

The even Gauss map $T_e : [0, 1] \to [0, 1]$ (Fig. 7.3) is given by

$$T_e(x) = \begin{cases} \dfrac{1}{x} - 2k & \text{if } x \in \left(\dfrac{1}{2k+1}, \dfrac{1}{2k} \right] \\ \dfrac{-1}{x} + 2k & \text{if } x \in \left(\dfrac{1}{2k}, \dfrac{1}{2k-1} \right] \\ 0 & \text{if } x = 0. \end{cases} \quad k \geq 1$$

If we plug in the even continued fraction expansion to T_e, we again delete the first digit (a_1, e_1) of x, i.e.,

$$T_e\left([\![(a_1, e_1)(a_2, e_2)(a_3, e_3)\ldots]\!]\right) = [\![(a_2, e_2)(a_3, e_3)(a_4, e_4), \ldots]\!].$$

That is, $(a_1, e_1) = (2k, +1)$ when $x \in \left(\frac{1}{2k+1}, \frac{1}{2k}\right]$ and $(a_1, e_1) = (2k, -1)$ when $x \in \left(\frac{1}{2k}, \frac{1}{2k-1}\right]$.

In order to construct the natural extension of the even Gauss map, we need to define the *extended even continued fractions.*[3] For $y \in [-1, 1]$, the extended even continued fraction expansion is

$$y = \cfrac{f_0}{b_0 + \cfrac{f_1}{b_1 + \cfrac{f_2}{b_2 + \cdots}}} = \langle\!\langle (f_0/b_0)(f_1/b_1)(f_2/b_2)\ldots \rangle\!\rangle,$$

While we will not define the Gauss map for the extended even continued fractions here, we can look at the first digit (f_0/b_0) get a sense of the difference between the two continued fraction expansions. For the *extended even continued fractions,* $f_0 = +1$ if $y \in (0, 1]$ and $f_0 = -1$ if $y \in [-1, 0)$. Then, $b_0 = 2k$ for $|y| \in (\frac{1}{2k-1}, \frac{1}{2k+1})$. For the *even continued fractions,* in the above notation, $f_1 = +1$ if b_0 is an underestimate for $\frac{1}{x}$ and $f_1 = -1$ if b_0 is an overestimate for $\frac{1}{x}$. The difference in notation emphasizes that the method for determining, and therefore indexing, the digits is different. This difference also appears in the way the natural extension moves the digits (Fig. 7.4).

We now define $\bar{T}_e$ on $[0, 1] \times [-1, 1]$ to be

$$\bar{T}_e(x, y) = \begin{cases} \left(\dfrac{1}{x} - 2k, \dfrac{1}{2k + y}\right) & \text{if} \quad x \in \left(\dfrac{1}{2k + 1}, \dfrac{1}{2k}\right] \\ \left(\dfrac{-1}{x} + 2k, \dfrac{-1}{2k + y}\right) & \text{if} \quad x \in \left(\dfrac{1}{2k}, \dfrac{1}{2k - 1}\right] \\ (0, y) & \text{if} \quad x = 0. \end{cases}$$

Looking at the even continued fraction expansion of $x = [\![(a_1, e_1)(a_2, e_2), \ldots]\!]$ and the extended even continued fraction expansion of $y = \langle\!\langle (f_0/b_0)(f_1/b_1), \ldots \rangle\!\rangle$, we get

[3] It is possible to define a natural extension of the even Gauss map without the dual continued fractions, but the invariant measure then depends on whether $e_1 = +1$ or $e_1 = -1$. The geometric construction would also need to be more complicated, involving cases for $e_1 = \pm 1$. The regularly continued fractions are what we call *self-dual,* meaning the dual continued fractions are still the regular continued fractions.

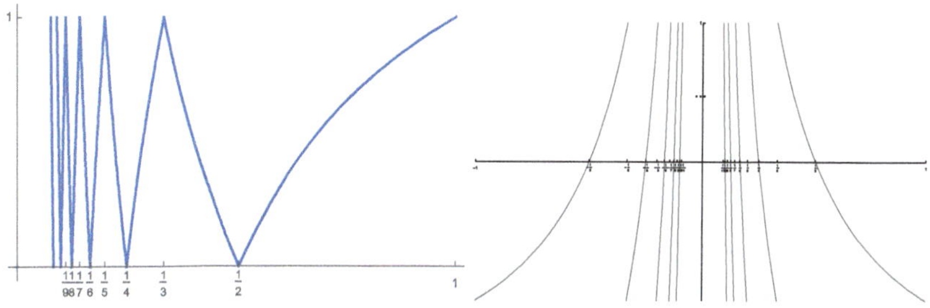

Fig. 7.4 Compare the graph of the even Gauss map (left) with the extended even Gauss map (right). The even Gauss map is a "folded" version of the extended even Gauss map

$$(x, y) = \left(\cfrac{1}{a_1 + \cfrac{e_1}{a_2 + \dots}}, \cfrac{f_0}{b_0 + \cfrac{f_1}{b_1 + \dots}} \right),$$

$$\overline{T}_e(x, y) = \left(\cfrac{1}{a_2 + \cfrac{e_2}{a_3 + \dots}}, \cfrac{e_1}{a_1 + \cfrac{f_0}{b_0 + \dots}} \right).$$

Thus, $\overline{T}_e$ moves the first digit of x to the first digit of y. In fact, $\overline{T}_e^{-1}$ is the natural extension of the extended even Gauss map, so we only need to study $\overline{T}_e$ to describe both the even continued fractions and the extended even continued fractions.

7.4 Hyperbolic Space

In order to connect continued fractions to geometry, we use the hyperbolic plane. One way to think about it is a space with too much material around each point to flatten out, like lettuce or coral. In the usual plane, six equilateral triangles can share a vertex without gaps or overlaps. A sphere can have up to five equilateral triangles sharing a vertex, like an inflated icosahedron. The hyperbolic plane can have up to infinitely many equilateral triangles sharing a vertex without gaps or overlaps.

We need a way to draw hyperbolic space, so we define the *upper half plane* where $\mathbb{H} := \{x+iy \mid y > 0\} \cup \{\infty\}$. We call the "top" of the upper half plane ∞. In order to distinguish the upper half plane from the Euclidean plane, we have to define a new notion of distance. In $\mathbb{H}$, the shortest path between two points, called a *geodesic*, is on a semicircle centered on the x-axis or on a straight line perpendicular to the x-axis.

A *hyperbolic triangle* is a region of upper half plane bounded by three geodesics, like how a Euclidean triangle is bounded by three straight lines. We say that two vertical lines have a vertex ∞ at the "top" of the upper half plane, so form two sides

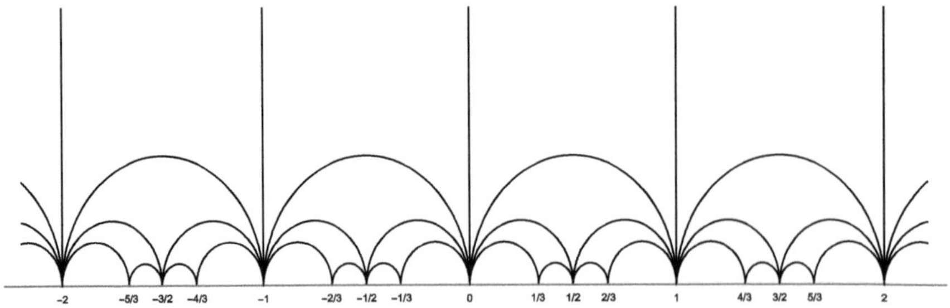

Fig. 7.5 Part of the Farey tessellation with geodesics connecting rational numbers with denominators less than or equal to 3

of a hyperbolic triangle. When I need to refer to a specific hyperbolic triangle, I will list the vertices.

In Fig. 7.5, there are three hyperbolic triangles drawn for each of the vertices $-3/2, -1/2, 1/2$, and $3/2$. There are also six hyperbolic triangles drawn for at $-2, -1, 0, 1$, and 2. The six triangles that share 0 as a vertex are as follows: $\{-1/2, -1/3, 0\}$, $\{-1, -1/2, 0\}$, $\{-1, 0, \infty\}$, $\{0, 1, \infty\}$, $\{0, 1/2, 1\}$, and $\{1/3, 1/2, 0\}$.

7.4.1 *Farey Tessellation and Hyperbolic Geodesics*

In order to describe continued fractions using geometry, we also need a *tessellation* or tiling of the upper half plane. In this case, each "tile" is a hyperbolic triangle with vertices either on the x-axis or at ∞. A tessellation is a rule for covering a space with tiles with no gaps or overlaps. In two dimensions, a tessellation is often called a tiling, like how tiles create patterns that cover a floor or wall without any gaps or overlaps.

To generate continued fractions using hyperbolic geometry, we use the *Farey tessellation*. First, we draw a vertical line at each integer. Next, for each n we define a set $F_n = \{\frac{p}{q} : p, q \in \mathbb{Z}, 0 < q \leq n\}$ which is the set of rational numbers with denominator less than or equal to n. We will list the numbers in order.

$$F_1 = \{\ldots, -2, -1, 0, 1, 2, \ldots\}$$
$$F_2 = \{\ldots, -\tfrac{3}{2}, -1, -\tfrac{1}{2}, 0, \tfrac{1}{2}, 1, \tfrac{3}{2}, \ldots\}$$
$$F_3 = \{\ldots, -1, -\tfrac{2}{3}, -\tfrac{1}{2}, -\tfrac{1}{3}, 0, \tfrac{1}{3}, \tfrac{1}{2}, \tfrac{2}{3}, 1, \tfrac{4}{3}, \ldots\}$$

To finish drawing the Farey tessellation, we connect each number to its neighbors in each F_n using a geodesic. So for F_1, we connect each integer k to $k-1$ and $k+1$. For F_2, we connect $k + 1/2$ to k and $k+1$. Figure 7.5 shows the tessellation up to F_3.

Fig. 7.6 Some oriented geodesics crossing two sides of a hyperbolic triangle, which meet at a vertex on the left

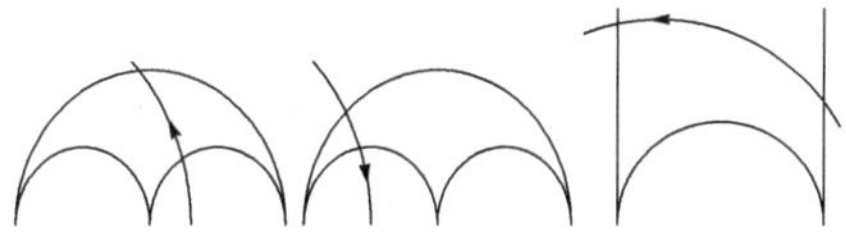

Fig. 7.7 Some oriented geodesics crossing two sides of a hyperbolic triangle, which meet at a vertex on the right

For the geometric description of continued fractions, we will consider *oriented geodesics*, which have a starting point and an ending point on the x-axis. We will walk along a geodesic γ from the starting point γ_- to the ending point γ_+ to find continued fraction expansions of γ_- and γ_+ (Figs. 7.6 and 7.7).

7.4.2 *Cutting Sequences and Continued Fractions*

The geometric description of the regular continued fractions given here is largely due to Series' paper "The modular surface and continued fractions" [14]. In addition to the argument outlined below, she uses tangent bundles, the topology of the infinite strings, and some group theory. The next few sections outline her argument but leave out details in order to focus on the big ideas.

We can use an oriented geodesic to find the regular continued fraction expansion of its starting and ending points based on how it crosses tiles in the Farey tessellation. Consider an oriented geodesic γ starting at $\gamma_- \in (0, 1)$ and ending at $\gamma_+ < -1$, as in Fig. 7.8, or starting at $\gamma_- \in (0, 1)$ and ending at $\gamma_+ < -1$ as in Fig. 7.9. The geodesic crosses a sequence of tiles in the Farey tessellation, each time the crossing two sides of the hyperbolic triangle. Since γ must cross the line $x = 0$, we will label the point $\gamma \cap \{x = 0\} = u_\gamma$. The part of the geodesics between γ_- and u_γ will help us find the continued fraction expansion of γ_-, while the part of the geodesic between u_γ and γ_+ will help us find the continued fraction expansion of γ_+.

Let us start with the case $\gamma_- \in (-1, 0)$, $\gamma_+ > 1$, as in Fig. 7.8. When $\gamma_+ \in [n, n+1)$, the geodesic γ crosses the n vertical lines $x = 1, 2, \ldots, n$ after u_γ. In Fig. 7.8, $\gamma_+ = \alpha \in [2, 3)$, and γ crosses the vertical lines $x = 1$ and $x = 2$. We mark the point where γ crosses $x = n$ with v_γ. Another way of saying this is γ crosses n Farey cells between u_γ and v_γ. For each cell, γ crosses two edges which "meet" at infinity. If you walk along the geodesic from γ_- to γ_+, infinity will be on

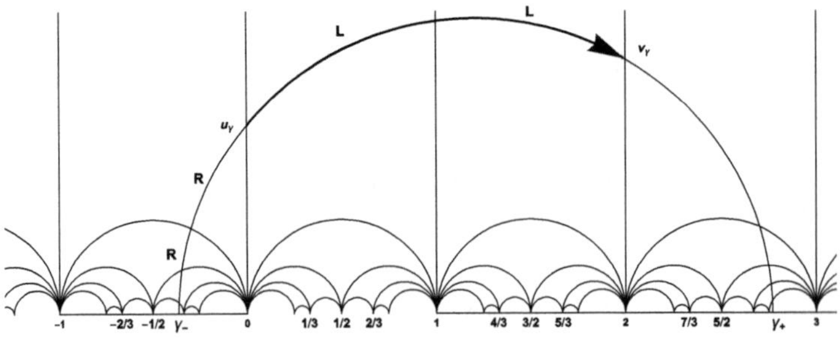

Fig. 7.8 The oriented geodesic starting at $\gamma_- = -1/\alpha$ and ending at $\gamma_+ = \alpha$. Segments are labeled L or R based on how the oriented geodesic crosses the Farey tessellation, giving the cutting sequence $R^2 u_\gamma L^2 v_\gamma RL$

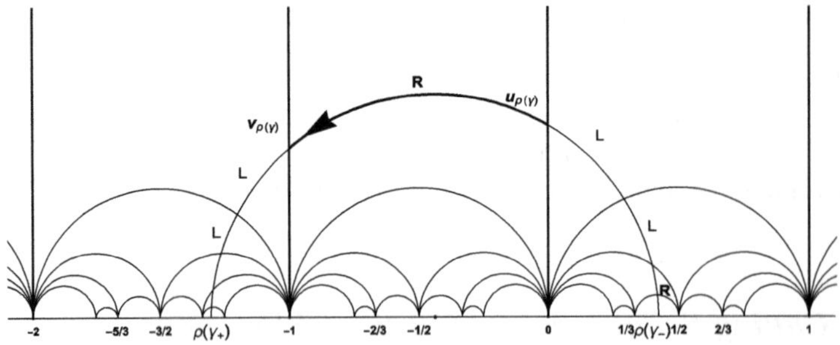

Fig. 7.9 The oriented geodesic connecting $\rho(\gamma_+)$ and $\rho(\gamma_-)$ when $\gamma_+ = \alpha$ and $\gamma_- = -1/\alpha$. Segments are labeled L or R based on how the oriented geodesic crosses the Farey tessellation giving the cutting sequence $RL^2 u_{\rho(\gamma)} R v_{\rho(\gamma)} L^2$

your left as you cross each of the n cells between u_γ and v_γ. Thus, we label each of the n segments of γ between u_γ and v_γ with L. When $\gamma_- \in (0, 1)$ and $\gamma_+ < -1$, infinity is on the right, so the segments are labeled R, as in Fig. 7.9.

Since it is less clear how to get the rest of the continued fraction expansions of γ_- and γ_+, the next step is to define a map that acts like $\overline{T}$ on (γ_+, γ_-) to find the next digit. Let $\gamma_+ = \operatorname{sign}(\gamma_+)[n_1; n_2, n_3, \ldots]$ and $\gamma_- = -\operatorname{sign}(\gamma_+)[0; n_0, n_{-1}, n_{-2}, \ldots]$. Define

$$\rho(x, y) = \left(\frac{1}{\operatorname{sign}(\gamma_+)n_1 - x}, \frac{1}{\operatorname{sign}(\gamma_+)n_1 - y}\right). \tag{7.7}$$

For $\gamma_+ = \alpha$ and $\gamma_- = -1/\alpha$, $\rho(\gamma_+, \gamma_-) = \left(\frac{1}{2-\left(2+\frac{1}{1+\ldots}\right)}, \frac{1}{2+\frac{1}{\alpha}}\right) = \left(-\left(1+\frac{1}{3+\ldots}\right), \frac{1}{2+\frac{1}{\alpha}}\right)$.

In general, ρ takes the first digit of γ_+ and moves it to the first digit of γ_-, just like $\overline{T}$. It also multiplies both endpoints by -1, but that does not cause any problems. By abuse of notation, I will also use $\rho(x)$ to refer to the image of one endpoint and $\rho(\gamma)$ for the oriented geodesic connecting $\rho(\gamma_+, \gamma_-)$. The section of the cutting sequence between u_γ and v_γ describes first digit of the regular continued fraction expansion of γ_+, the section of the cutting sequence between $u_{\rho(\gamma)} = \rho(v_\gamma)$ and $v_\rho(\gamma)$ describes the second digit, and so on. We can use these ideas to find the continued fraction expansion of γ_+, like how we can use T to find the regular continued fraction expansion of a number.

Theorem 7.2 (Series' Theorem A [14]) *For the oriented geodesics described above:*

- *When* $\gamma_+ = [n_1; n_2, \ldots]$ *and* $\gamma_- = -[0; n_0, n_{-1}, \ldots]$, *the geodesic has cutting sequence* $\ldots L^{n_{-1}} R^{n_0} u_\gamma L^{n_1} v_\gamma R^{n_2} \ldots$.
- *When* $\gamma_+ = -[n_1; n_2, \ldots]$ *and* $\gamma_- = [0; n_0, n_{-1}, \ldots]$, *the geodesic has cutting sequence*
  $\ldots R^{n_{-1}} L^{n_0} u_\gamma R^{n_1} v_\gamma L^{n_2} \ldots$.

There are a few things to notice before sketching a proof for our example:

$$\gamma_+ = [2; 1, 3, 3, \ldots], \quad \gamma_- = -[0; 2, 1, 3, 3, \ldots].$$

The first is that the cutting sequences of γ_+ and γ_- are independent. That is, if we take the two geodesics γ from γ_- to γ_+ and γ' from γ'_- to γ_+, the cutting sequence describing γ_+ is the same. The second thing to notice is that if one endpoint is rational, there is ambiguity in the cutting sequence, since the last letter could be L or R. This corresponds to the ambiguity of the continued fraction expansion, where $L^n R$ corresponds to $n + \frac{1}{1} = n + 1$, which in turn corresponds to L^{n+1}. The ambiguity of cutting sequences is why the rational case is trickier (Fig. 7.10).

Proof (Proof of Theorem 7.2 for $\gamma_+ = \alpha$ ***and*** $\gamma_- = \frac{-1}{\alpha}$***)*** The cutting between $x = 0$ and γ_+: Let v_γ be the last place that γ crosses a vertical line $x = n_1$. In this case, v_γ is where γ crosses the line $x = 2$. Then the bold section of γ in Fig. 7.8 crosses 2 tiles between u_γ and v_γ, both labeled L. Thus, we know part of the cutting sequence is $u_\gamma L^2 v_\gamma$. Any oriented geodesic ending between 2 and 3 will cross the line $x = 2$ and then turn right to cross the semicircle connecting 2 to 3. Thus, the

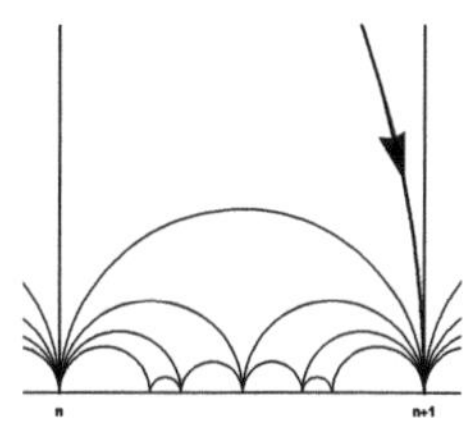

Fig. 7.10 A geodesic ending at the corner $n + 1$, which could be labeled L or R

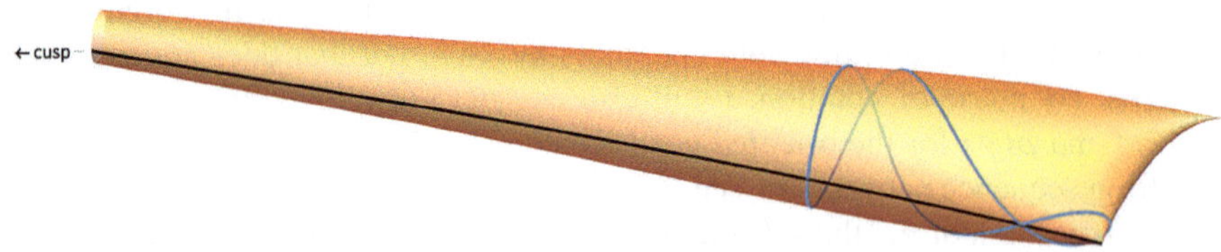

Fig. 7.11 A geodesic on the modular surface which is a closed loop

cutting sequence will contain $u_\gamma L^2 v_\gamma$, and the first digit of the continued fraction expansion is 2.

To find the next part of the sequence, we need to look at $\rho(\gamma_+, \gamma_-)$ since ρ deletes the first digit of the regular continued fraction expansion of γ_+ and moves it to γ_-.

The geodesic $\rho(\gamma)$ is shown in Fig. 7.9. Note that $\rho(v_\gamma) = u_{\rho(\gamma)}$. We define $v_{\rho(\gamma)}$ similarly to above, but now we have $\rho(\gamma_+) = -[1; 3, 3, \ldots]$. Thus, $v_{\rho(\gamma)}$ is where $\rho(\gamma)$ crosses $x = -1$, and $\rho(\gamma)$ crosses one triangle labeled R. So, now we know that part of the cutting sequence is $u_\gamma L^2 v_\gamma R^1 v_{\rho(\gamma)}$. Continuing this process will give the entire cutting sequence for geodesics that end at γ_+.

The cutting sequence between γ_- and $x = 0$: We use $\frac{-1}{\gamma_-}$ to convert to the above cases. The action $\frac{-1}{z}$ flips Ls to Rs and vice versa. However, we have already found the cutting sequence for $\frac{-1}{\gamma_-} = \gamma_+$ starts with $L^2 R^1$. Thus, we now know that our example cutting sequence looks like $\ldots L^1 R^2 u_\gamma L^2 v_\gamma R^1 \ldots$, agreeing with the labeling in Fig. 7.8. □

The cutting sequences corresponding to regular continued fractions allow us to describe geodesics on the corresponding modular surface. There is a map which projects the hyperbolic upper half plane $\mathbb{H}$ onto the surface in Fig. 7.11, where the "cusp" end continues on to infinity, and the black line is the projection of the edges of the Farey tessellation. Oriented geodesics on $\mathbb{H}$ are mapped to oriented geodesics on the surface with the same cutting sequence. On the surface, the cutting sequence L means the oriented geodesic starts at the black line with the cusp on the left, wraps around the surface, and comes back to the black line. The meaning of R is analogous.

Some geodesics on the modular surface start and end at the same point, like the one in Fig. 7.11. You can use cutting sequences to prove that these geodesics are the images of oriented geodesics on the hyperbolic surface where $\gamma_+ = \overline{[a_0; a_1, \ldots, a_{2r}]}$ and $\gamma_- = -[0; \overline{a_{2r}, a_{2r-1}, \ldots, a_1}]$, where the line means the digits repeat, as in repeating decimals. The continued fraction expansion of the endpoints then provides a (somewhat complicated) formula for the length of the geodesic.

We can also use ρ to help us find the invariant measure for $\overline{T}$. First, remember that ρ takes the set $[1, \infty) \times [-1, 0] \cup (-\infty, -1] \times [0, 1]$ to itself. Next, we define a map $f(x, y)$ from $[1, \infty) \times [-1, 0] \cup (\infty, -1] \times [0, 1]$ to $[0, 1]^2$ by

$$f(x, y) = \begin{cases} \left(\frac{1}{x}, -y\right) & \text{if } x \geq 1 \\ \left(\frac{-1}{x}, y\right) & \text{if } x \leq -1 \end{cases} = \text{sign}(x)\left(\frac{1}{x}, -y\right),$$

and $\overline{T} \circ f = f \circ \rho$. The map f is onto except for the line $(0, y)$ which has no preimage.

Proposition 7.1 *The invariant measure for* ρ *is* $\iint h(x, y)dxdy = \iint \frac{dxdy}{(x-y)^2}$ *(see, for example, Series [14, Section 3]).*

We will use the invariant measure of ρ to find the invariant measure of $\overline{T}$. To simplify the notation, let $S = [1, \infty) \times [-1, 0] \cup (\infty, -1] \times [0, 1]$. The change of variables formula from multivariable calculus lets us use f^{-1} to find the invariant measure for $\overline{T}$.

Since f is not invertible, we have two cases: $f^{-1}(u, v) = (\frac{1}{u}, -v)$ and $f^{-1}(u, v) = (\frac{-1}{u}, v)$. In the first case, $f^{-1}(u, v) = (\frac{1}{u}, -v) = (x, y)$, so $x = \frac{1}{u}, y = -v$, and the Jacobian is $\frac{1}{u^2}$. Finally, the change of variables formula gives

$$\iint_S h(x, y)dxdy = \iint_{[0,1]^2} \frac{h(f^{-1}(u, v))}{u^2} dudv = \iint_{[0,1]^2} \frac{1}{u^2} \frac{dudv}{(\frac{1}{u} - (-v))^2}$$
$$= \iint_{[0,1]^2} \frac{dudv}{(1 + uv)^2}.$$

The other case gives the same result. Thus, the invariant measure for $\overline{T}$ is $\iint \frac{dxdy}{(1+xy)^2}$.

It is possible to directly verify that $\frac{dxdy}{(1+xy)^2}$ is the invariant measure for $\overline{T}$, as [3] does for the even continued fractions. However, the geometric construction allowed us to find the measure more easily.

7.5 Contributions

Along with my advisor Florin Boca, I described a cutting sequence representation of even continued fractions, as well as those with odd denominators [3]. I have also described the slow continued fractions, where every digit is $(1, +1)$ or $(2, -1)$, corresponding to the same surface as the regular continued fractions [3]. More recently, I have been using the cutting sequence in this section, and the even continued fractions, to study a map which appears while generating Pythagorean triples [12].

For the even continued fractions, we translate from the cutting sequence for regular continued fractions to the cutting sequence for even continued fractions according to the insertion and singularization algorithm.

The cutting sequences of regular continued fractions and even continued fractions both use the Farey tessellation, but they have different underlying geometries as described in the introduction. (See references [3, 14] for more details and [12, Section 3] Section 3 for an alternate coding.) Each of the continued fraction expansions corresponds to a specific group and provides a description of something called *group equivalence*. The correspondence between the groups and continued fractions are another result that we get from the geometric construction.

7.5.1 Even Continued Fractions

We can also use cutting sequences to find the even continued fraction expansion of γ_+ and the extended even continued fraction expansion of γ_-. We will use the same last step as the regular continued fractions, that is, for some set $S_e \subset \mathbb{R}^2$, the map $f : S_e \to [0, 1] \times [-1, 1]$ is given by

$$f(x, y) = \begin{cases} \left(\frac{1}{x}, -y\right) & \text{if } x \geq 1 \\ \left(\frac{-1}{x}, y\right) & \text{if } x \leq -1 \end{cases} = \operatorname{sign}(x) \left(\frac{1}{x}, -y\right),$$

and $\overline{T_e} \circ f = f \circ \rho_e$. We use $[0, 1] \times [-1, 1]$ for this map since $\overline{T_e}$ is defined on $[0, 1] \times [-1, 1]$. Now, we just need to find S_e, ρ_e, and a way to convert between cutting sequences and the even and extended even continued fraction expansions of γ_+ and γ_-. Note that since f is the same as in the regular continued fraction case, the multivariable calculus change of variables formula gives that the invariant measure for $\overline{T_e}$ is again $\iint \frac{dxdy}{(1+xy)^2}$.

Proposition 7.2 *Since we are still working with the Farey tessellation, the invariant measure for* ρ_e *is still* $\iint \frac{dxdy}{(x-y)^2}$. *Since we are using the same function* f *as in Proposition 7.1, the invariant measure for* $\overline{T}_e$ *is still* $\iint \frac{dxdy}{(1+xy)^2}$.

Again, it is possible to directly verify that $\iint \frac{dxdy}{(1+xy)^2}$ is the invariant measure for $\overline{T}_e$ (for example, Section 6.1 of [3]).

To find S_e, we can use $f^{-1}(x, y) = (\gamma_+, \gamma_-)$. Then the two cases are

$$(\gamma_+, \gamma_-) = \left(\frac{1}{x}, -y\right) \in [1, \infty) \times [-1, 1],$$

$$(\gamma_+, \gamma_-) = \left(\frac{-1}{x}, y\right) \in (-\infty, -1] \times [-1, 1].$$

That is, we consider the cases $\gamma_+ \geq 1$ and $\gamma_+ \leq -1$. In both cases, $\gamma_- \in [-1, 1]$.

Since these geodesics will no longer cross $x = 0$, the lines $x = 1$ and $x = -1$ will separate the part of γ corresponding to γ_+ and γ_-. In this case, γ will cross $x = 1$ when γ_+ is positive, and it will cross $x = -1$ when γ_+ is negative. To account for this change, u_γ is now the place γ crosses $x = \pm 1$ and v_γ is the next time γ crosses an arc connecting two integers. Since v_γ is on an arc between integers, $n, n+1$, we know $n < \gamma_+ < n+1$. The cutting sequence between u_γ and v_γ will describe the first digit (ordered pair (a_i, e_i)) of the even continued fraction expansion of γ_+. This way, we move both u_γ and v_γ one letter to the right in the cutting sequence.

From Eq. (7.6), we know that

$$\begin{aligned}
\alpha &= [2; 1, 3, 3, \dots] = 2 + \cfrac{1}{2 - \cfrac{1}{2 - \cfrac{1}{2 - \cfrac{1}{4+z}}}} \\
&= [\![(2, +1); (2, -1)(2, -1)(2, -1)(4, +1), \dots]\!], \\
\frac{1}{\alpha} &= [0; 2, 1, 3, 3, \dots] = \cfrac{1}{2 + \cfrac{1}{2 - \cfrac{1}{2 - \cfrac{1}{2 - \cfrac{1}{4+z}}}}} \\
&= \langle\!\langle (+1/2)(+1/2)(-1/2)(-1/2)(-1/4), \dots \rangle\!\rangle.
\end{aligned} \tag{7.8}$$

From Figs. 7.8 and 7.9 and the process described in Sect. 7.4.2, we know that the cutting sequence for the geodesic with endpoints $\gamma_+ = \alpha$ and $\gamma_- = \frac{-1}{\alpha}$ contains $L^3R^3L^1R^2L^2R^1L^3R^3$, where the imaginary axis is between $L^3R^3L^1R^2$ and $L^2R^1L^3R^3$. We can also see these strings in Figs. 7.8 and 7.12, which show the geodesic from $\gamma_- = \frac{-1}{\alpha}$ to $\gamma_+ = \alpha$ along with its cutting sequence. We rewrite the cutting sequence to include the new u_γ and v_γ and get $L^3R^3L^1R^2Lu_\gamma LRv_\gamma L^3R^3$. As we will show below, the first "digit" (ordered pair) of the even continued fraction expansion of γ_+ is determined by the portion of the cutting sequence between u_γ and v_γ. In the case of α, the first digit is $(2, +1)$; we see that this corresponds to the sequence LR.

When $\gamma_+ \in [2k, 2k+1)$, we know from the regular continued fraction case that the cutting sequence for γ is $\dots Lu_\gamma L^{2k-1}Rv_\gamma \dots$, so $L^{2k-1}R$ corresponds to the digit $(2k, +1)$. This matches with the example where LR corresponds to $(2, +1)$. When $\gamma_+ \in [2k-1, 2k)$, the cutting sequence is $\dots Lu_\gamma L^{2k-2}Rv_\gamma \dots$, and $L^{2k-2}R$ corresponds to the digit $(2, -1)$. Again, the case $\gamma_+ \leq -1$ is analogous,

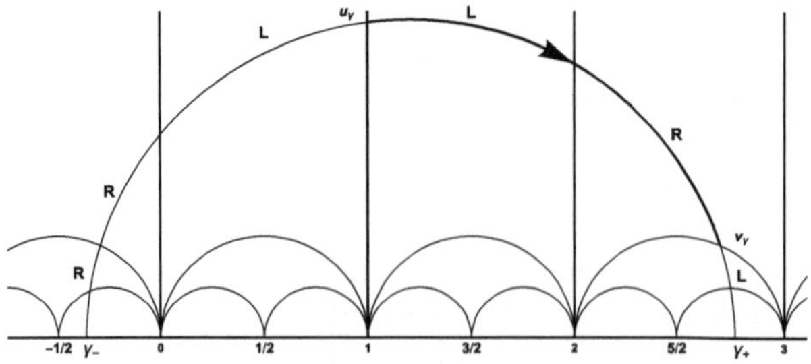

Fig. 7.12 The geodesic γ with endpoints $\gamma_+ = \alpha$ and $\gamma_- = -1/\alpha$. Segments are labeled L or R based on how the geodesic crosses the Farey tessellation giving the sequence $R^2 L u_\gamma L R v_\gamma L$

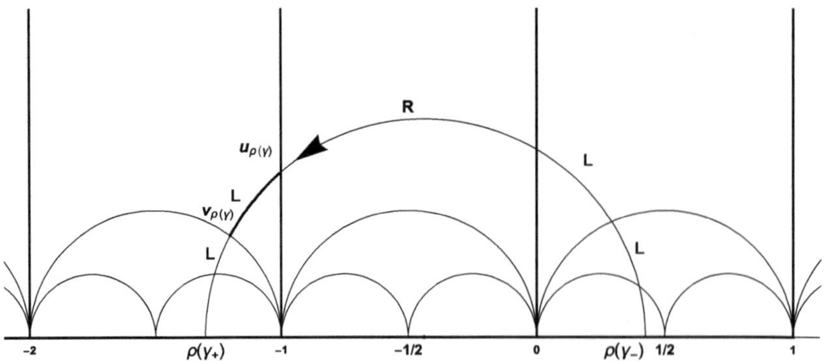

Fig. 7.13 The geodesic $\rho_e(\gamma)$ with endpoints $\rho(\gamma_+)$ and $\rho_e(\gamma_-)$. Segments are labeled L or R based on how the geodesic crosses the Farey tessellation giving the sequence $RL^2 R u_{\rho(\gamma)} L v_{\rho(\gamma)} L$

with the roles of L and R switched. Note that it is possible to get $L^0 R$ when $\gamma_+ \in [1, 2)$ and $R^0 L$ when $\gamma_+ \in (-2, -1]$, as in Fig. 7.13.

As with the regular continued fractions, we define a map ρ_e that acts like $\bar{T}_e$ on (γ_+, γ_-). That is, ρ_e takes the first digit of γ_+ and moves it to the first digit of γ_-. Thus, the first digit of the even continued fraction expansion corresponds to the section of the cutting sequence between u_γ and v_γ. The second digit corresponds to the section of $\rho_e(\gamma)$ between $u_{\rho_e(\gamma)}$ and $v_{\rho_e(\gamma)}$, and we can continue this process to find the even continued fraction expansion of γ_+.

If

$$\gamma_+ = \operatorname{sign}(\gamma_+) \left(a_1 + \cfrac{e_1}{a_2 + \cfrac{e_2}{a_3 + \cfrac{e_3}{\dots}}} \right) = \operatorname{sign}(\gamma_+) [\![(a_1, e_1); (a_2, e_2)(a_3, e_3), \dots]\!],$$

$$\gamma_- = -\operatorname{sign}(\gamma_+) \left(\cfrac{e_0}{a_0 + \cfrac{e_{-1}}{a_{-1} + \cfrac{e_{-2}}{a_{-1}} + \dots}} \right)$$

$$= -\text{sign}(\gamma_+)\langle\!\langle (e_0/a_0, e_0)(e_{-1}/a_{-1})(e_{-2}/a_{-2}), \dots \rangle\!\rangle,$$

we define

$$\rho_e(x, y) = \left(\frac{1}{\text{sign}(\gamma_+)a_1 - x}, \frac{1}{\text{sign}(\gamma_+)a_1 - y} \right)$$

as in the regular continued fraction case. Now

$$\begin{aligned}
&\rho_e(\gamma_+, \gamma_1) \\
&= \left(\frac{1}{\text{sign}(\gamma_+)a_1 - x}, \frac{1}{\text{sign}(\gamma_+)a_1 - y} \right) \\
&= \left(\frac{1}{\text{sign}(\gamma_+)a_1 - \text{sign}(\gamma_+)\left(a_1 + \cfrac{e_1}{a_2 + \cfrac{e_2}{\dots}}\right)}, \frac{1}{\text{sign}(\gamma_+)a_1 + \text{sign}(\gamma_+)\left(\cfrac{e_0}{a_0 + \dots}\right)} \right) \\
&= \left(-\text{sign}(\gamma_+)e_1\left(a_2 + \frac{e_2}{\dots}\right), \frac{\text{sign}(\gamma_+)}{a_1 + \cfrac{e_0}{a_0 + \dots}} \right) \\
&= \left(-\text{sign}(\gamma_+)e_1\left(a_2 + \frac{e_2}{\dots}\right), \text{sign}(\gamma_+)e_1 \frac{e_1}{a_1 + \cfrac{e_0}{a_0 + \dots}} \right).
\end{aligned}$$

When $e_1 = +1$, we multiply by -1 as in the regular continued fraction case, but do not when $e_1 = -1$. Again, we find that $\rho_e(v_\gamma) = u_{\rho_e(\gamma)}$.

Returning to our example $\gamma_+ = \alpha$, $\gamma_- = -1/\alpha$, we see that

$$\begin{aligned}
\rho_e(\gamma_+, \gamma_-) = (&[\![(2, -1); (2, -1)(2, -1)(4, +1), \dots]\!], \\
&\langle\!\langle (+1/2)(+1/2)(+1/2)(-1/2)(-1/2), \dots \rangle\!\rangle).
\end{aligned}$$

The cutting sequence of $\rho_e(\gamma)$ contains $L^3R^3L^1R^2Lu_\gamma LRu_{\rho_e(\gamma)}Lv_{\rho_e(\gamma)}L^2R^3$. Thus, $u_{\rho_e(\gamma)}Lv_{\rho_e(\gamma)}$ corresponds to $(2, -1)$ as desired. Since we cannot (easily) tell from the cutting sequence when there is a hidden L^0 or R^0, summarizing the rules for reading the even continued fraction expansion of γ_+ from the cutting sequence is more complicated. See [12, Section 3] for a simplified description of the cutting sequence, which uses a modified version of the tessellation.

Fig. 7.14 The modular surface corresponding to the even continued fractions, with a geodesic shown in blue

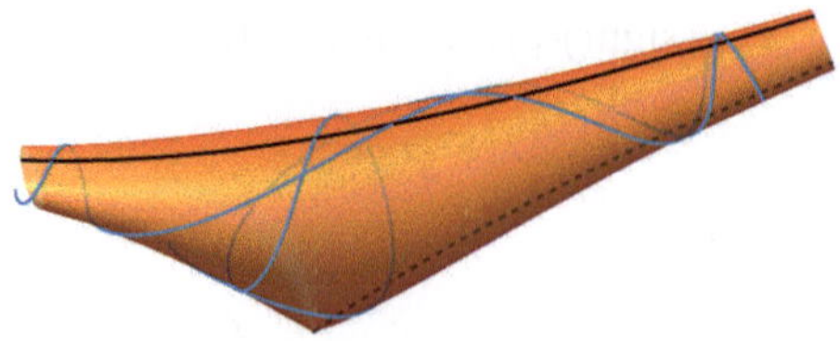

Theorem 7.3

- *For a geodesic with* $\gamma_+ > 1$ *and* $-1 < \gamma_- < 1$, *we find:*
  - $*$ $\gamma_+ = [\![(2k_1, +1); (2k_2, e_2)\dots]\!], \gamma_- = -\langle\!\langle(2k_0, e_0)(2k_{-1}, e_{-1})\dots\rangle\!\rangle$ *has cutting sequence* $\dots Lu_\gamma L^{2k_1-1} Rv_\gamma R\dots$.
  - $*$ $\gamma_+ = [\![(2k_1, -1); (2k_2, e_2)\dots]\!], \gamma_- = -\langle\!\langle(2k_0, e_0)(2k_{-1}, e_{-1})\dots\rangle\!\rangle$ *has cutting sequence* $\dots Lu_\gamma L^{2k_1-2} Rv_\gamma R\dots$ *when* $2k_1 > 2$ *and* $\dots Lu_\gamma L^0 Rv_\gamma R\dots = \dots Lu_\gamma Rv_\gamma R\dots$ *when* $2k_1 = 2$.

  We use $\rho_e(\gamma)$ *to find the next digit string corresponding to* $(2k_2, e_2)$.
- *For a geodesic with* $\gamma_+ < -1$ *and* $-1 < \gamma_- < 1$, *we find:*
  - $*$ $\gamma_+ = -[\![(2k_1, +1); (2k_2, e_2)\dots]\!], \gamma_- = \langle\!\langle(2k_0, e_0)(2k_{-1}, e_{-1})\dots\rangle\!\rangle_e$ *has cutting sequence* $\dots Ru_\gamma R^{2k_1-1} Lv_\gamma L\dots$.
  - $*$ $\gamma_+ = -[\![(2k_1, -1); (2k_2, e_2)\dots]\!], \gamma_- = \langle\!\langle(2k_0, e_0)(2k_{-1}, e_{-1})\dots\rangle\!\rangle_e$ *has cutting sequence* $\dots Ru_\gamma R^{2k_1-2} Lv_\gamma L\dots$ *when* $2k_1 > 2$ *and* $\dots Ru_\gamma Lv_\gamma L\dots$ *when* $2k_1 = 2$.

  We use $\rho_e(\gamma)$ *to find the next digit string corresponding to* $(2k_2, e_2)$.
- *To find the first digit of the extended even continued expansion of* γ_-, *we apply* ρ_e^{-1} *and read the cutting sequence between* $u_{\rho_e^{-1}(\gamma)}$ *and* $v_{\rho_e^{-1}(\gamma)}$ *using the same rules. We continue this process to get the rest of the digits.*

Finally, we have a similar relationship between the even continued fraction expansion of γ_+, extended even continued fraction expansions of γ_-, and geodesics on the surface in Fig. 7.14. The black line is the image of the lines $x = 2k + 1$, and the dashed line is the image of the lines $x = 2k$, with the other edges of the Farey tessellation projecting to either the black line or the dashed line. Since there are two cusps, the relationship between the cutting sequence and the cusps is more complicated. However, we can again prove that geodesics which start and end at the same point are the images of oriented geodesics on the hyperbolic surface where $\gamma_+ = [\![\overline{(a_0, e_1); (a_1, e_1)\dots, (a_r, e_r)}]$ and $\gamma_- = -[0; \overline{(e_r/a_r), e_{r-1}/a_{r-1}), \dots,)e_1/a_1)}]$, as well as a formula for the length of the geodesics.

7.6 Ideas for Future Study

There are many online resources with basic patterns with continued fractions. The website [9] has a long list of patterns in the digits of regular continued fraction expansions. The article "Christiaan Huygens' Planetarium" [2] proves several number theoretic facts about regular continued fractions before describing how Huygens used continued fractions to construct a mechanical planetarium.

If you are interested in learning more about geometric constructions of continued fractions, it would be good to take number theory, dynamics, and differential geometry courses. The textbook *Dynamics Done with Your Bare Hands* [4] introduces several ways of studying dynamical systems. The first chapter is the most similar to this chapter and is available online [5]. On the continued fraction side, Masarotto [10] carefully describes many types of continued fraction expansions in her master's thesis. More closely related to the topic in this chapter is Svetlana Katok's course notes *Continued fractions, hyperbolic geometry and quadratic forms* [7], which use the *minus* continued fractions (where every numerator is -1). The book *Ergodic Theory of Numbers* is intended for an advance undergraduate audience and goes more in depth on the connections between continued fraction expansions and probability theory.

My paper "Coding of geodesics on some modular surfaces and applications to odd and even continued fractions" with Florin Boca [3] discusses the cutting sequences corresponding to even continued fractions in greater depth. This paper also describes the cutting sequences corresponding to the continued fractions with odd denominators which relate to a different subgroup of determinant 1 matrices. For both the even and the odd continued fractions, the paper describes some of the number theoretic results, as well as using the continued fraction expansions to calculate lengths of geodesics on the surfaces.

Once you are familiar with tangent spaces and some topology, I suggest reading Series' paper [14] that was outlined in Sect. 7.4.2. This paper carefully describes the regular continued fraction coding using ideas from differential geometry and topology.

For graduate-level sources, Katok and Ugarcovini [8] and Adler and Flatto [1] wrote articles on many connections between hyperbolic geometry and continued fractions. Adler and Flatto in particular discuss the connections between the hyperbolic plane and matrix groups. Finally, Einsiedler and Ward's book *Ergodic theory with a view towards Number Theory* [6] covers many of related topics from dynamical systems.

Acknowledgments I am grateful to the SMP community for their constant support through graduate school, including while writing this chapter. I would also like to thank the referees for their suggestions that greatly improved this chapter. This research was completed while I was a graduate student at the University of Illinois at Urbana-Champagin.

References

1. Adler, R.L., Flatto, L.: Cross section maps for geodesic flows. I. The modular surface. In: Ergodic Theory and Dynamical Systems, II (College Park, Md., 1979/1980). Progress in Mathematics, vol. 21, pp. 103–161. Birkhäuser, Boston (1982)
2. Amin, H.: Christiaan Huygens' planetarium. http://www.irem.univ-mrs.fr/IMG/pdf/huygens-delft.pdf (2008)
3. Boca, F.P., Merriman, C.: Coding of geodesics on some modular surfaces and applications to odd and even continued fractions. Indagationes Math. **29**(5), 1214–1234 (2018). https://doi.org/10.1016/j.indag.2018.05.004, http://www.sciencedirect.com/science/article/pii/S0019357718301988
4. Dal'Bo, F., Ledrappier, F., Wilkinson, A. (eds.): Dynamics done with your bare hands. EMS Series of Lectures in Mathematics. European Mathematical Society (EMS), Zürich (2016). Lecture notes by Diana Davis, Bryce Weaver, Roland K. W. Roeder, Pablo Lessa, Lectures given at the Undergraduate Summer School "Boundaries and Dynamics" held at the University of Notre Dame, Notre Dame, IN (2015)
5. Davis, D.: Lines in positive genus: An introduction to flat surfaces. Preprint. arXiv:1507.02571 (2015)
6. Einsiedler, M., Ward, T.: Ergodic Theory with a View Towards Number Theory. Springer, Berlin (2013)
7. Katok, S.: Continued fractions, hyperbolic geometry and quadratic forms. In: Katok, S., Sossinsky, A., Tabachnikov, S. (eds.) MASS Selecta: Teaching and Learning Advanced Undergraduate Mathematics, pp. 121–160. American Mathematical Society, Providence (2003)
8. Katok, S., Ugarcovici, I.: Symbolic Dynamics for the Modular Surface and Beyond, vol. 44, pp. 87–132 (2007)
9. Knott, R.: Introduction to continued fractions. https://r-knott.surrey.ac.uk/fibonacci/CFintro.html
10. Masarotto, V.: Metric and arithmetic properties of a new class of continued fraction expansions. Ph.D. Thesis, Master Thesis, Universita di Padova and Leiden University (2009)
11. Merriman, C.: Geometric and ergodic properties of certain classes of continued fractions. PhD Thesis, University of Illinois at Urbana-Champaign (2019)
12. Merriman, C.: Geodesic flows and slow down maps of continued fraction maps. Preprint. arXiv:2307.11700.pdf (2023)
13. Schweiger, F.: Continued fractions with odd and even partial quotients. Arbeitsber. Math. Inst. Univ. Salzburg **4**, 59–70 (1982)
14. Series, C.: The modular surface and continued fractions. J. London Math. Soc. **2**(1), 69–80 (1985)

Chapter 8
Stochastic Models in Biology

Deena R. Schmidt

Abstract Mathematical biology is an interdisciplinary field that lies at the interface of mathematics and biology. Mathematics plays an important role at all levels of biological organization and regulation. My research is driven by a desire to understand the roles of stochasticity (noise), structure, and evolution in shaping the dynamics of biological systems. I develop and analyze mathematical models, combining methods from probability and statistics, dynamical systems, and random graph theory, to shed light on biological issues while generating new mathematical questions. Stochasticity is fundamental to gene expression and ion channel gating in neurons; it also plays an underappreciated role in physiological regulation at larger scales, such as the regulation of sleep and wake behavior. In my research, I investigate the effects of stochasticity, as well as physiological structure (such as neuronal networks), and the constraints of evolution, on real-world biological dynamics. A particular challenge in this work is the need to bridge multiple spatial and temporal scales. Below I provide an overview of the field of mathematical biology and I describe my contributions to this field. I use methods such as stochastic models in genetics and stochastic processes on graphs to understand interlocking small and large facets of biological regulation.

Keywords Mathematical biology · Stochastic model · Markov chain · Population genetics · Wright-Fisher model · Moran model · Waiting time distributions · Gene regulatory sequence evolution · Multi-stage carcinogenesis · Networks · Random graphs · Stochastic processes on networks · Neuronal networks · Neural systems

---

D. R. Schmidt (✉)
University of Nevada, Reno, NV, USA
e-mail: drschmidt@unr.edu

A. Brisbin et al. (eds.), *Research Connections*, Association for Women in Mathematics Series 36, https://doi.org/10.1007/978-3-031-81217-0_8

8.1 Biography/Career Trajectory

Ever since I can remember, math and science have intrigued me, challenged me, and forced me to think in different ways. I was the weird kid who actually enjoyed solving story problems in elementary school, loved AP calculus and AP biology in high school, and took upper-level math and statistics courses for fun in college and graduate school. In college, abstract mathematics opened my eyes and my mind, leading me to pursue two outstanding summer undergraduate research programs: Carleton and St. Olaf Summer Mathematics Program for Women (SMP) and the Mathematical and Theoretical Biology Institute (MTBI) held at Cornell University. These amazing research experiences are what led me to graduate school. I dabbled in both pure and applied mathematics while an undergrad, initially declaring a double major (Mathematics and Dance) and a possible third major (Biology), but ultimately deciding to major in pure Math with minors in Biology and Dance. I then chose to study applied math in grad school, as I had become fascinated by the integration of mathematics and biology at MTBI. As a PhD student in the Center for Applied Mathematics (CAM) at Cornell, I studied mathematical biology, probability, and stochastic processes and took a few courses in genetics and neuroscience to build up my biological knowledge. I chose Cornell because of the interdisciplinary focus of their applied math program, the individualized course of study, and the "gorge"ous location.

Choosing to study mathematical biology was a no-brainer for me, but focusing on a specialty within the large field of math bio took some time. For my first two years at Cornell, I was funded by a National Science Foundation (NSF) training grant fellowship in mathematical biology which required that I take a sequence of courses on mathematical ecology, topics in theoretical epidemiology, and a seminar course. As I talked with other CAM graduate students about my research plans, I would mention my interest in genetics applications. Several students suggested that I talk to Rick Durrett; I followed their advice and started working with Durrett in my second year. Durrett is an expert in probability, stochastic processes, and biological applications. During the summer between my second and third years, I worked as a teaching assistant for Durrett's Research Experience for Undergraduates (REU) on probability problems arising from genetics. This experience gave me a good sense of how to conduct research in this field.

In my third year, I received an NSF Graduate Research Fellowship which helped me focus my time and energy on research. My PhD dissertation looks at developing and analyzing models of gene regulatory sequence evolution. Specifically, I used probabilistic models to understand how population genetic factors influence this evolutionary process and shape the distribution of waiting times for the generation of new transcription factor binding sites in different species. This work and extensions of this work are described in more detail in the Contributions section.

Upon graduation from Cornell, I applied for three postdoctoral positions and received offers from two of them: the Institute for Applied Mathematics and Its Applications (IMA) at the University of Minnesota and the Mathematical

Biosciences Institute (MBI) at Ohio State University. I accepted the offer from the IMA and deferred my offer to the MBI for one year. I was very fortunate and excited to have these fantastic postdoctoral opportunities. I spent one year at the IMA and participated in their thematic program on the mathematics of molecular and cellular biology. I spent the following three years at the MBI and loved every minute of it. With an average of eight workshops held at these institutes per year, I interacted with many influential researchers in mathematical biology; I continued learning new mathematical tools and application areas. It was at the MBI that I became interested in network theory and neuroscience applications while working with Janet Best, an accomplished mathematical neuroscientist.

At the end of my MBI postdoc, my husband was just finishing his PhD (we met as graduate students at Cornell and we both work in mathematical biology). He was offered an MBI postdoctoral fellowship so I found a third postdoc position at Case Western Reserve University (located in Cleveland, OH, about two hours north of Ohio State) to temporarily solve our two-body problem. I worked with Peter Thomas, a mathematical neuroscientist who also works with stochastic models. It was a great fit and my most productive postdoc of the three, although they were all great experiences. I was funded by an NSF grant and my focus was on theoretical analysis of biological networks. In particular, I studied how stochastic (or noisy) perturbations of activity on a given network affect measurable variables capturing important aspects of network activity and how this activity might be influenced by network structure. Networks are fundamental objects of interest throughout biology, and I developed a theoretical framework which has applications to a wide variety of biological settings.

Currently, I am an Associate Professor in the Department of Mathematics and Statistics at the University of Nevada, Reno. Academically, I am in a great place. My university values both teaching and research and is supportive of faculty with kids. My department has a vibrant group of applied mathematicians, probabilists, and statisticians that I interact with daily; they form a strong network of support for me and the other junior faculty in the department. The biology department is located in the building adjacent to ours which allows for many opportunities to interact with biologists. I have several new collaborations with biologists and neuroscientists here at UNR. Lastly, I am a mother of three: an 11 year-old boy and 7-year-old identical twin girls. I have written an article on balancing an academic career and family life [31].

My advice to current undergraduate students who are considering graduate school is to first figure out what you are passionate about and why. Graduate school can be a hard and long experience, but if you enjoy the field you study and the research you conduct, it does not necessarily feel that way. There are several things you can do now to figure this out. Talk to graduate students or professors at your institution about research in different areas. Take a challenging class in the area you are considering. Participate in an undergraduate research experience before applying to graduate schools. Undergraduate research gives you the chance to work independently on a problem that likely does not have a known solution, a very different perspective than just taking classes. If you have the opportunity, attend a research conference with your mentor and present your results. I recently brought a new undergraduate researcher to a Society for Industrial and Applied Mathematics (SIAM) conference and she was absolutely invigorated by the experience, eager to dive into her own research project.

As for graduate students, you hopefully already figured out what you are passionate about and why; if not, I highly recommend seeking out various mentors (these can be more senior graduate students) to help you figure that out sooner rather than later. Use this knowledge to stay inspired and to propel yourself from one stage to the next. Ask for help when you need it. When you are ready to apply for jobs, know that the job market fluctuates and you have to be resilient. Trust in your abilities.

8.2 Introduction

The field of mathematical biology is interdisciplinary, lying at the interface of mathematics and biology. This field is sometimes also called biomathematics and focuses on the use of mathematical tools to study biological systems. To quote Lee Segal, a distinguished applied mathematician and one of the founders of modern mathematical biology, "mathematical biology sounds like a narrow specialty, but in fact it encompasses all of biology and most of mathematics." Mathematics plays an important role at all levels of biological organization and regulation. In addition, biology is currently undergoing a quantitative revolution with the advancement of genome science, incredible technological advances, and the rapid growth of huge data sets. Mathematical biologists are in demand to make sense of these big data sets (via analysis, modeling, informing new experiments, informing public health policies, etc.) which makes it an exciting time to work in this field.

Research in mathematical biology is truly interdisciplinary. It involves talking with biologists to become familiar with their experiments and data with respect to a specific biological process. We develop a mathematical model that describes the biological process, which can be done in different ways. For instance, we could use ordinary or partial differential equations, or we could use stochastic models which possess some inherent randomness (or depend on chance events). Another important step is to simulate the biological process and compare the numerical results with experimental results. This often continues iteratively by revising the model until the fit to experimental results is sufficient. Finally, we use the model to make new hypotheses and suggest new experiments.

The broad field of mathematical biology includes many areas of specialized research, such as quantitative genetics and evolutionary biology (which includes population genetics, coalescent theory, and phylogenetics), modeling cellular and molecular biology (e.g., cancer modeling and simulation, modeling of the cell cycle, modeling of intracellular dynamics), mathematical epidemiology (the study of infectious diseases), and computational neuroscience (the theoretical study of the nervous system), just to name a few. Within the broad list of research topics above,

I work on problems in applied stochastic processes with a focus on applications in genetics and neuroscience.

Population genetic models look at how various evolutionary forces such as random genetic drift, mutation, migration, and selection change the frequency at which new gene types (called alleles) appear within a population over time [15]. These processes all have an inherent element of chance or randomness, making stochastic models ideal for their study. In a stochastic model, the same set of parameter values and initial conditions will lead to an ensemble of different outputs, or, in other words, each output is drawn from a probability distribution of possible outcomes. In contrast, a deterministic model, such as a set of ordinary differential equations, predicts a single outcome for a given input and thus cannot model the inherent randomness of many real-world situations; there is no randomness in deterministic models. Below I will formally define a stochastic process and give examples of how such a model is used in applications.

In probability and statistics, a random variable is a variable whose possible values depend on the outcomes of a random experiment or phenomenon. A stochastic process, then, is a collection of random variables $\{X(t) : t \in T\}$ where T is an index set (usually thought of as time) [26]. For each $t \in T$, $X(t)$ is a random variable representing the state of the system at time t. The state space S describes all possible values the random variable can take on, and S can be discrete or continuous. Likewise, the index set T can be discrete or continuous, but we will focus on the discrete-time case here. Typically, a discrete-time stochastic process uses time $T = \mathbb{N}_0$, the set of natural numbers including 0. In this way, a stochastic process is used to model the evolution of some physical process over time.

As a simple example, suppose $T = \mathbb{N}_0$ and $S = \{1, 2, 3\}$. This is a discrete-time and discrete-space stochastic process. Let $X(0) = 1$ define the initial condition, meaning that the stochastic process starts in state 1 at time 0. From state 1, the process can move to either state 2 or state 3 with equal probability in the next time step, and this process continues on (moving to one of the other 2 states in the next step with equal probability). We can visualize this process as a network of three nodes (states 1, 2, and 3) with directed edges between the nodes representing possible transitions between states and the associated edge weights describing the transition probabilities. See Fig. 8.1 for an illustration of the network diagram.

A sample path, also called a realization, of the 3-state process is as follows:

$$X(0) = 1,\ X(1) = 2,\ X(2) = 3,\ X(3) = 2,\ X(4) = 3,\ X(5) = 2,\ X(6) = 1\ldots.$$

In Fig. 8.2, we show an illustration of the sample path described above for the 3-state example. The realization shows the change in state (black dots) over time for 25 discrete-time steps, starting with the process in state 1 at time 0. This type of stochastic process is also known as a Markov chain. A Markov chain satisfies a property known as the Markov property, which says that the future state of the process only depends on the current state and not on any past history. This property is mathematically very useful in making complicated stochastic processes analytically tractable.

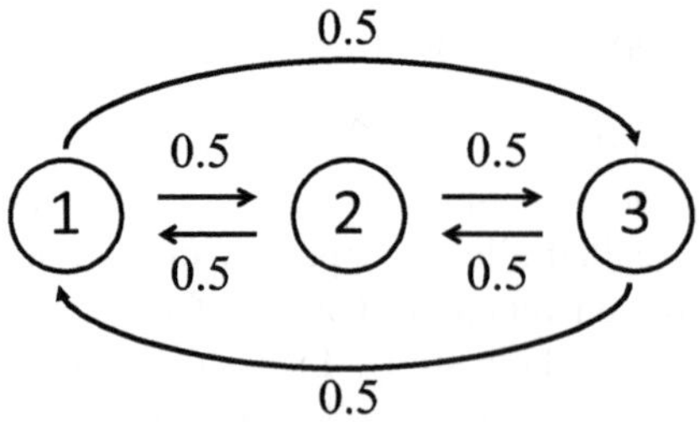

Fig. 8.1 Network diagram for the 3-state stochastic process example $\{X(t) : t \in \mathbb{N}_0\}$ where $S = \{1, 2, 3\}$; this stochastic process is also a Markov chain. The transition probabilities between states are given by the edge weights (equivalently, see matrix P below)

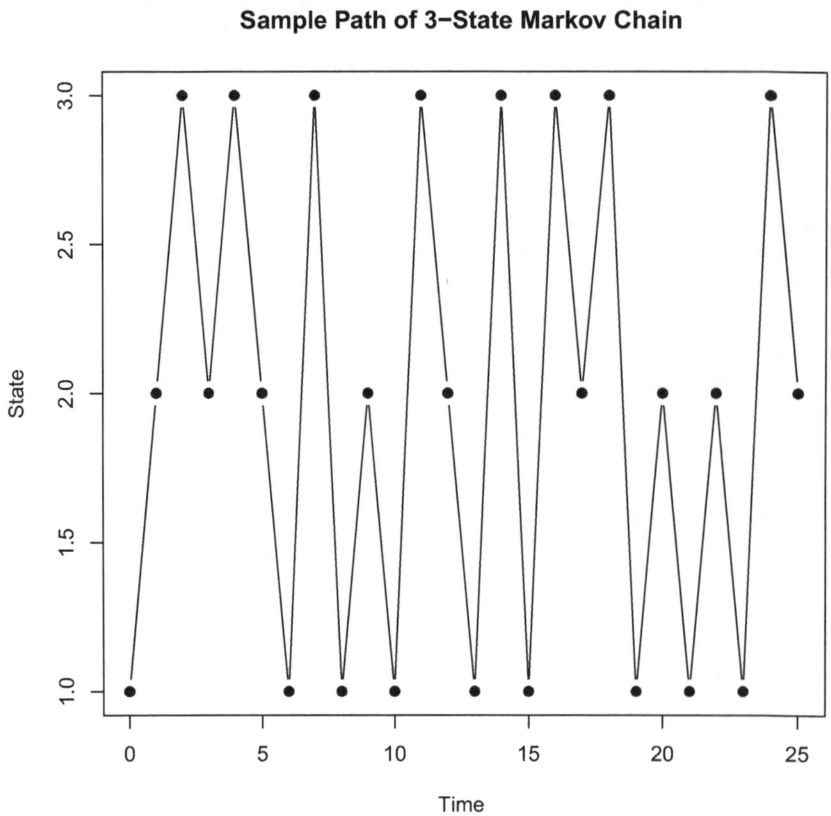

Fig. 8.2 Realization of the 3-state Markov chain example $\{X(t) : t \in \mathbb{N}_0\}$ where $S = \{1, 2, 3\}$ and transition probabilities between states given by matrix P (equivalently, see the edge weights in Fig. 8.1). This sample path has initial condition $X(0) = 1$ and is shown for 25 discrete-time steps

The transition probability matrix $P = (P_{ij})_{i,j \in S}$ of a Markov chain specifies how the process moves forward in time by defining conditional probabilities:

$$P_{ij} = P(X(t+1) = j \mid X(t) = i) \text{ for all } i, j \in S \text{ and } t = 0, 1, 2, \ldots.$$

In words, P_{ij} is the probability that, given that the process is in state i at time t (the current state), the process moves to state j in the next time step $t+1$. For the simple example above, we have that

$$P_{ij} = P(X(t+1) = j \mid X(t) = i) = \begin{cases} 0.5 & \text{if } i \neq j \in S \\ 0 & \text{if } i = j \in S \end{cases}$$

which gives us the following transition probability matrix:

$$P = \begin{pmatrix} 0 & 0.5 & 0.5 \\ 0.5 & 0 & 0.5 \\ 0.5 & 0.5 & 0 \end{pmatrix}.$$

For instance, we interpret entry $P_{1,2} = 0.5$ (row 1, column 2) as the one-step transition from state 1 to state 2 occurring with probability 0.5. Note that the entries on the diagonal of P are all 0 because this Markov chain does not allow transitions from a state back to itself; such transitions are known as self-loops. The reason this stochastic process is also a Markov chain is because the transition probabilities P_{ij} are independent of the past history. In mathematical terms,

$$P(X(t+1)=j \mid X(t) = i)=P(X(t+1)=j \mid X(t) = i, \ldots, X(1) = i_1, X(0) = i_0)$$

for $i_0, i_1, \ldots, i, j \in S$.

Now let us consider a biological example. More complicated versions of this model will be used in the next section. Let A and a represent two different alleles, or gene types, of a particular gene in a population with N individuals. Note that in diploid populations (such as humans and many other animals and plants), each individual has two copies of each gene which means that the population has $2N$ copies of each gene. In contrast, haploid populations (such as human egg and sperm cells) only have one copy of each gene. The *Wright-Fisher model* in population genetics is a discrete-time and discrete-state stochastic process describing the evolution of the number of type A alleles in the population over time (measured in generations). As a concrete example, think of a diploid annual plant population that reproduces with one generation per year. Let $X(t)$ denote the number of A alleles in the population at generation t. The state space S is the set of possible counts of type A alleles in the population, i.e., $S = \{0, 1, \ldots, 2N\}$. At each generation, a collection of alleles is sampled with replacement from the current population at generation t to form the new generation at time $t+1$. Hence, this process describes binomial sampling of alleles each generation. In other words, the probability of transitioning from i type A alleles in the population at generation t to j type A alleles in the population at generation $t + 1$ is given by the binomial probability distribution. More specifically, we use the binomial probability mass function with size $2N$ (total number of alleles) and success probability equal to the frequency $x = i/2N$ of allele A in the population at generation t to compute the transition probability:

$$P_{ij} = P(X(t+1) = j \mid X(t) = i) = \binom{2N}{j} x^j (1-x)^{2N-j}.$$

This quantity, P_{ij}, is an element in the probability transition matrix $P = (P_{ij})_{i,j \in S}$ which specifies how this process evolves forward in time.

Recall that the Wright-Fisher model is a model for animal/plant populations (e.g., insects, annual plants) that reproduce in discrete generations. Humans and many other animals reproduce continuously so this model is not appropriate for such populations. A version of this model which allows for overlapping generations, known as the *Moran model*, is more appropriate. Both the Wright-Fisher model and the Moran model are Markov chains and discrete-state stochastic processes. The Wright-Fisher model has discrete-time steps (generations), whereas the Moran model has continuous time steps (overlapping generations). The latter is modeled as a birth and death process, a particular type of continuous-time Markov chain. For those who would like to know more, [15, 22] contain good discussions of both models.

I use Markov chains extensively in my research, as described in the next section. I am particularly interested in the interplay between stochasticity (noise) and network structure in biological processes, as well as the influence of evolutionary forces in shaping the dynamics of biological systems. The work described below uses methods such as stochastic models in genetics and stochastic processes on networks (graphs) to gain insight into different types of biological regulation. I also briefly describe a few projects that focus on fitting models to data.

8.3 Contributions

8.3.1 Stochastic Models in Genetics

My work on stochastic modeling in genetics uses mathematical models to explore how genes can regulate macroscopic behavior and how the DNA itself changes. Estimating the amount of time for an event or a sequence of events to occur is a fundamental problem in probability theory, commonly modeled by a Poisson process or other type of stochastic process. Waiting time models are used to study many problems in genetics, including understanding the mechanisms involved in gene regulation, the speed at which regulatory changes can occur, and the rate of cancer onset in multistage carcinogenesis. To gain insight into these complicated problems, I have used a variety of tools from probability theory (such as Markov chains) to develop and analyze models of stochastic gene regulation which yield new insights into evolutionary genetics, cancer progression, and related diseases, as described below. A central difficulty in such problems lies in capturing the essential components of the process in a mathematical model amenable to analysis, computation, and prediction.

8.3.1.1 Waiting Time Distributions: Gene Regulatory Sequence Evolution

Genetics is the study of genes, genetic variation, and heredity in organisms. A gene is a sequence of DNA that codes for a protein that has a particular function, and the transmission of genes to an organism's offspring are the basis of trait inheritance. The building blocks of DNA are called nucleotides and they come in four different types: adenine (A), cytosine (C), guanine (G), and thymine (T). A 10-letter DNA sequence, for example, could be CCAGTCAATT.

A regulatory sequence is a short sequence of DNA (in vertebrates many are 6–10 nucleotides long) located near a gene where regulatory proteins known as transcription factors (TF) bind to control the expression of that gene. Essentially, the expression level of a gene is how much gene product is produced, which can influence the strength of the gene's effect. Figure 8.3 gives an illustration of a 6-nucleotide regulatory sequence AACCGT which functions as a binding site for transcription factors. The binding site is located near a gene (represented by the large rectangle). Mutations, or changes in nucleotides, within regulatory sequences can cause changes in gene expression which in turn can lead to phenotypic evolution. The phenotype is an organism's observable characteristics or traits such as its physical form, developmental processes, and behavior. For instance, humans and chimpanzees have different phenotypes (they look different) even though their DNA sequences are quite similar, with DNA sequence similarity estimated to be more than 98% identical [14].

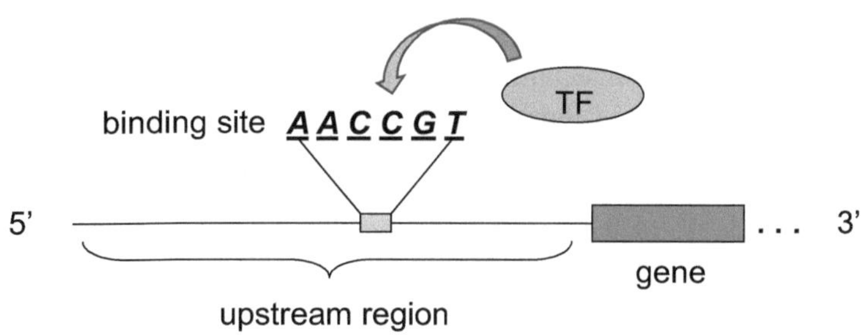

Fig. 8.3 Illustration of a gene and its upstream regulatory region which is where regulatory sequences are found. Here, a 6-nucleotide regulatory sequence AACCGT functions as a binding site for transcription factors (TF)

One possible explanation for the substantial differences between humans and chimpanzees is that there have been extensive changes in regulatory sequences, rather than in genes themselves. This raises quantitative questions about the rate of regulatory sequence evolution in humans and whether it is sufficiently rapid to contribute to the differences between humans and chimps. Using this biological framework, I considered the following probability problem: given a gene regulatory region of length L in the human genome, how long does it take for a word of length W to appear by random mutation in that region in some individual?

I analyzed this problem using a variety of Markov chain models, in collaboration with Rick Durrett (Duke Univ.) [7, 30]. We built up to the problem described above by considering three simpler models. I will describe the first model in detail and then briefly discuss the other two. In the first model, we fixed a target word of length W, say AACCGT, a 6-letter regulatory sequence. We then considered a discrete-time Markov chain $\{X(n) : n \in \mathbb{N}_0\}$ such that $X(n)$ changes each time there is a mutation (change in letter) and gives the number of letters that match the W letter target sequence. The state space for this Markov chain is $S = \{0, 1, \ldots, W\}$. For example, $X(n) = W$ means that all letters match the target word (in the correct order), whereas $X(n) = 2$ means that only 2 letters match the target sequence. We focused on target words of length $W = 6$ and $W = 8$ which are common lengths for DNA regulatory sequences. The goal of the first model is to describe the waiting time distribution for a given W-letter target word to appear by mutation in a W-letter sequence with random initial condition. This Markov chain has transition probabilities:

$$\begin{aligned} P_{i,i-1} &= \frac{i}{W} \\ P_{i,i+1} &= \left(\frac{1}{3}\right)\left(\frac{W-i}{W}\right) \\ P_{i,i} &= \left(\frac{2}{3}\right)\left(\frac{W-i}{W}\right) \end{aligned} \tag{8.1}$$

where all other $P_{i,j} = 0$. The state decreases by 1 if we choose a matching letter to mutate. If we choose a nonmatching letter to mutate, the state only increases by 1 if we choose the correct one of three possible mutations which will match the target word. We assume that all mutations are equally likely.

What we are really interested in is the distribution of the first time the Markov chain reaches state W, or the first time a match to the target word is reached. In probability terms, this is represented by

$$T_W = \inf\{t : X(t) = W\}, \tag{8.2}$$

where inf stands for *infimum* and refers to the greatest lower bound of this set [27]. The infimum is similar to the minimum, but more general in that it allows one to characterize sets with no minimum. The distribution of T_W is known as a *waiting time distribution.* Notice the change in index set for the Markov chain in Definition 8.2. We use $t \in [0, \infty)$ instead of $n \in \mathcal{N}_0$ because here we consider a continuous-time Markov chain (CTMC) $\{X(t) : t \in [0, \infty)\}$ that has the same sequence of jumps (state transitions) as the discrete-time Markov chain (DTMC) $\{X(n) : n \in \mathbb{N}_0\}$ described above. The only difference between these two Markov chains is the times at which the jumps occur, with the CTMC being the more appropriate model for studying waiting times. The times between jumps in the CTMC model follow an exponential distribution with a particular rate. The CTMC is a more realistic model because it allows for variable amounts of time between jumps rather than assuming jumps happen every discrete-time step as in the DTMC.

In the second model, we considered waiting for a target W-letter word to appear somewhere within a sequence of L nucleotides. While L is general, we are thinking about L as 1 or 2 thousand nucleotides which is biologically reasonable for a regulatory region containing transcription factor binding sites. Specifically, we use a continuous-time Markov chain that jumps according to an exponential distribution. When a jump occurs, a randomly chosen letter is changed to a different randomly chosen letter. In this case, we again want to quantify the waiting time distribution.

The third model essentially puts the first model into a population context. The population consists of N "individuals," modeled as W-letter words. We use the continuous-time Moran model to describe the dynamics of the population and how it evolves forward in time. The Moran model is a stochastic process commonly used in population genetics modeling [21, 22]. Let the population consist of N individuals, and let μ be the mutation rate of each letter in the sequence. We suppose that each individual lives for an exponentially distributed amount of time (with mean 1) and then is replaced. To replace an individual, we choose at random from the whole set of individuals, including itself, and make a copy. With probability $W\mu$ we randomly change one of the letters in the W-letter sequence, and the new individual joins the population. After N such replacement steps, we say that 1 generation has passed. If we start with a population of W-letter words that do not match the prespecified target word (e.g., AACCGT), how many generations does it take for an individual in the population to match the target word? How long does it take for all individuals to match the target word? When a mutation becomes fixed in the entire population, this is called *fixation*. The last question could be rephrased to ask "what is the expected time to fixation for the target word?"

Finally, we can use this Moran model framework to put the second model into a population context and answer our main question: Given a gene regulatory region of length L in the human genome, how long does it take for a W-letter word to appear by random mutation in that region in some individual? We estimated average waiting times for the generation of words of length $W = 6$ and 8 within an $L = 1000$ nucleotide regulatory region in humans. We found that the waiting time follows a mixed exponential distribution that depends on the presence of an "almost match" in the population. For instance, in the case $W = 8$, if initially some individual matches 7 out of 8 letters to the target word, the average waiting time is significantly shorter than if we do not have that scenario. To see this, let X_1 and X_2 be exponential random variables with different expected values, ($E[X_1]$ and $E[X_2]$), and let f_{X_1}

and f_{X_2} be the corresponding probability density functions for random variables X_1 and X_2. Then we have the (approximate) mixture distribution describing the waiting time:

$$f_{T_W} \approx p_1 f_{X_1} + p_2 f_{X_2} \tag{8.3}$$

where p_1 and p_2 are probabilities that add to 1. For $W = 8$ in the human population, we have

$$f_{T_8} \approx \frac{5}{16} f_{X_1} + \frac{11}{16} f_{X_2} \tag{8.4}$$

where the expected values of X_1 and X_2 are $E[X_1] = 375{,}000$ years and $E[X_2] = 650$ million years, respectively. In this case, X_1 is the waiting time to fixation if initially some individual matches 7 out of 8 letters, and X_2 is the waiting time to fixation if this is not the case. Biologically, transcription factor binding often occurs with an inexact match, so if we stop the process after the first 7 out of 8 letter match appears in the population, then the expected value $E[T_8]$ drops to 60,000 years. We note that such flexibility in transcription factor binding is important to allow regulatory sequences to evolve at a reasonable rate in humans. Since we are focusing on differences between humans and chimpanzees, we compare the mean waiting time to the human-chimpanzee divergence time which has been estimated to be between five and seven million years ago [19]. Regulatory sequences should evolve faster than divergence times between species.

In [8], we extended the model described above and applied it to known divergence times between species for both humans and *Drosophila* (fruit flies). In particular, we examined the waiting time distribution for a pair of mutations: The first mutation (A) inactivates an existing transcription factor binding site and the second (B) creates a new active binding site within the regulatory region. This model is a general two-step process of binding site turnover where the mutations can occur in either order in an individual, but we focused on the sequence A then B. Using a continuous-time Moran model, we studied the time evolution of this mutation process under the effects of selection and random genetic drift in the human and *Drosophila* populations. We then quantified the mean waiting time for a type B mutation to arise in some individual, and also for the B mutation to become fixed in the population, under various fitness conditions for the A and B type mutation.

One of the key points of this study [8] was to look at how the tempo and mechanisms of regulatory sequence evolution depend on an organism's "effective population size." This is a basic parameter in population genetic models and is defined as the number of breeding individuals in an idealized constant size population that would be needed to mimic various properties of the real population, which may be growing or shrinking. We assumed a human effective population size of 10,000, as is commonly done. This value is relatively small compared to that of organisms such as *Drosophila* whose effective population size is in the millions. Applying our results, we found that a few million years is sufficient for the rate of binding site turnover in *Drosophila*, consistent with recent experimental observations, but in humans, this type of change would take more than 100 million years. Fortunately, as mentioned above, the match of a regulatory sequence to its target sequence does not have to be exact for binding to occur. In addition, there are at least 20,000 genes in the human genome, and tens or hundreds of pairs of mutations can occur for each one. Incorporating these details into the model significantly reduces the mean waiting time. Therefore, our findings show that regulatory sequences can evolve at rates faster than divergence times between species (since the human-chimpanzee divergence time is 5–7 million years ago [19]). Importantly, our results support the scientific explanation that changes in regulatory sequences are driving phenotypic evolution, specifically the differences we observe between humans and chimpanzees.

Our analysis also exposed some flaws in the literature concerning mathematical limits to Darwinian evolution [8, 9]. We used our results for the expected waiting time for a sequence of two mutations to investigate the emergence of chloroquine resistance in the parasite *Plasmodium falciparum* that causes malaria in humans. Genetic studies have shown that this is due to changes in a protein PfCRT and that in the mutant strains two amino acid changes are almost always present; see [40] for details. This example has been discussed in [4], and in [8, 9] we present a correction for the waiting time calculation for two such mutations to appear in one parasite, which is overestimated in [4] by more than 150,000 years.

8.3.1.2 Waiting Time Distributions: Multistage Carcinogenesis

Colon cancer and many other diseases are the result of a sequence of several mutations occurring in a collection of cells. I worked in collaboration with Jason Schweinsberg (UCSD) and Rick Durrett to generalize the waiting time results from the 2-step process described above by considering the time it takes for an individual cell to acquire m mutations in a population [10]. The case $m = 2$ applies to gene regulatory sequence evolution, but for cancer development, the number of mutations m may be greater than 2. Results from the general m-step model below give estimates for the rate of cancer progression in colon cancer and related diseases.

In [10], we considered the mathematical problem: Given a population of N individual cells, how long does it take until some individual cell has accumulated m specified mutations? We formulated a general m-step model that applies both to cancer development and to gene regulatory sequence evolution (for the case $m = 2$). We assumed that the population evolves according to the Moran model and that individual cells may acquire mutations during their lifetimes due to radiation or other environmental factors. We labeled each individual cell with a type $0 \leq j \leq m$ to denote the number of mutations acquired by that individual and assumed that individuals of type $j-1$ mutate to type j at rate u_j, given that initially all individuals are type 0. Let $X_j(t)$ be the number of type j individuals at time t. For each positive integer m, let

$$\tau_m = \inf\{t : X_m(t) > 0\} \tag{8.5}$$

be the first time at which there is an individual cell of type m in the population. Note that τ_1 is exponentially distributed with rate Nu_1. We computed the asymptotic distribution of τ_m for $m \geq 2$ as $N \to \infty$ under various conditions for how the mutation rates μ_j depend on the population size N.

An interesting case arises when the number of mutant cells is small relative to the total population size N. In that case, we can approximate the number of cells with mutations by a branching process in which each individual cell gives birth at rate 1 and dies at rate 1. A branching process is a particular type of continuous-time stochastic process used to model reproduction in biological systems [26]. Let the mutation rates μ_j be nonnegative functions of the population size N. For instance, we could have the following mutation rates: $\mu_1 = \dfrac{\lambda}{N}$ and $\mu_2 = \dfrac{1}{N}$ for some nonnegative constant, λ. Suppose that, in general, we have the following limiting conditions as $N \to \infty$:

1. $N\mu_1 \to \lambda \in [0, \infty)$, the product of the population size and mutation rate μ_1 is a constant, λ.
2. $\mu_2 \to 0$, the mutation rate μ_2 tends to 0.
3. $N\sqrt{\mu_2} \to \infty$, the product of the population size and mutation rate μ_2 tends to ∞.

Then we have the following asymptotic distribution of the (scaled) waiting time for an individual cell to acquire 2 mutations: $\tau_2 \cdot N\mu_1\sqrt{\mu_2}$ converges to a distribution that has a density function:

$$f_2(t) = h(t)\exp\left(-\int_0^t h(s)\right) \quad \text{where} \quad h(s) = \frac{1 - e^{-2s/\lambda}}{1 + e^{-2s/\lambda}} \tag{8.6}$$

if $\lambda > 0$ and $f_2(t) = \exp(-t)$ if $\lambda = 0$. This is Theorem 1 in [10] which is one of several cases considered for how the mutation rates μ_j scale with the population size N; the other cases modify the limiting conditions given above as $N \to \infty$. Complete results for general m along with rigorous proofs are given in this reference as well. We can apply these results to estimate rates of colon cancer progression by using the case $m = 4$ which assumes a well-known 4-stage model [20], or cases $m = 5, 6, 7$, or 8 which depend on whether or not the population has any inherited risk factors [5].

8.3.2 *Stochastic Processes on Networks*

A network is a collection of points (called vertices or nodes) connected together pairwise by links (called edges). Mathematically, a network is equivalent to a graph

$G = (V, E)$ which is defined by its set of vertices V and its set of edges E. An edge connecting vertex u to vertex v is defined as an ordered pair (u, v), but edges can be unordered as well. I use the terms network and graph interchangeably. Now suppose we generate and fix a network (graph). Then we can define a stochastic process to evolve on the nodes of the network, meaning that the state of each node in the graph can change over time but the physical network structure remains fixed. Network dynamics, then, refers to how the network nodes change state over time.

Networks with stochastic processes evolving on the nodes have become a very active area of research in mathematical biology, with applications in neuroscience, cell signaling, ecology, and other areas of computational biology. Many biological processes can be thought of as flows on networks, from ion channel dynamics in neuroscience to the flow of biomass through food webs in ecology. Network structure plays a key role in network dynamics, and finding reduced descriptions (lower dimensional approximations) of high-dimensional biological networks can help us to understand their function by uncovering their essential components. Most studies rely on numerical simulations or mean field[1] approximations of the underlying stochastic system. The research projects described below pair biologically-motivated numerical studies with mathematical analysis.

8.3.2.1 Network Structure and Dynamics of Biological Systems

In this section, I discuss modeling networks in which the structure of the network is not known with certainty, and in addition, there is a biologically inspired stochastic process occurring on the nodes of the network. Thus, there are (often different) probability distributions describing both the structure of the network and the process on the network.

A fundamental question regarding biological networks is the extent to which the network architecture may contribute to the dynamics occurring on the network. Recent studies have found interesting structure in data arising from biological systems, such as data following certain types of probability distributions. Several questions naturally arise: If trying to understand data produced by a process on a network, should we study the process or the network? How can we understand the contributions of the network topology and of the process itself in shaping the data? Explanations for how such structure arises sometimes focus on the processes generating the data and other times on the structure of the network.

In collaboration with Janet Best (OSU) and Mark Blumberg (Univ. of Iowa) [35], I explored the relative contributions to network dynamics made by the graph structure and by the nature of the stochastic process occurring on that graph. The

[1] In a stochastic system, state variable values change over time according to random events. The model defines a distribution for each point in the state space that describes where the process moves to over the next time step. A mean field approximation of such a model simply replaces that distribution with its mean. For an example mean field approximation, see [16, 17].

degree of a node is defined as the number of connections (via edges) it has to other nodes in the network, and the degree distribution is the probability distribution of these degrees over the whole network. Primarily using degree distribution as a marker of graph structure, we considered randomly generated graphs on N nodes with varied degree distributions. We focused on graphs with (approximately) Poisson and power law degree distributions; these degree distributions become exact as the number of nodes $N \to \infty$.

The Poisson distribution is a well-known probability distribution. A random variable X has a Poisson distribution with parameter $\lambda > 0$ if its probability mass function is given by

$$f(k) = P(X = k) = \frac{\lambda^k e^{-\lambda}}{k!} \quad \text{for } k = 0, 1, 2, \ldots. \tag{8.7}$$

A power law is a functional relationship between two quantities such that one quantity varies as a power of the other. The probability density function

$$f(x) = Cx^{-\alpha} \quad \text{for} \quad x \geq x_{\min} \tag{8.8}$$

has a power law relationship for large values of x, where C is a constant such that $\int_0^\infty f(x)dx = 1$ and $\alpha > 1$ is the power law exponent.

Figures 8.4 and 8.5 show networks generated to have these two different degree distributions: Poisson and power law. Figure 8.4 is an example of an Erdos-Renyi (ER) random graph [11, 12], one of the simplest types of random graph. Given a set of N nodes, a network is formed under the ER model by connecting each pair of nodes with probability p, independently of all other pairs of nodes in the network. In other words, a node is equally likely to be connected to any of its $N - 1$ neighbors. As a result, the degree distribution for an ER network is binomial with parameters $N - 1$ and p (recall that the binomial distribution approaches a Poisson distribution as $N \to \infty$).

The network shown in Fig. 8.5 is an example of a scale-free random graph, and it is generated by the preferential attachment model developed by Barabasi and Albert [3]. This model starts with a small number of connected nodes and adds nodes one by one until the network consists of N nodes. New connections are made to existing nodes with probability proportional to each node's degree, so that highly connected nodes are more likely to gain new connections. The right panel of Fig. 8.5 shows the corresponding degree distribution for this network. We see that most nodes have only one neighbor, but there are a few nodes that have many neighbors (up to 6 neighbors); this distribution approximately follows a power law relationship.

After generating networks with different structures, we then considered two different stochastic processes to put on these graphs: (1) an activation process and (2) a spiking process that models neurons firing in a neuronal network. The activation process assumes that nodes can be in one of two states, inactive or active. Initially, only a small number of nodes are active. At each discrete-time step, active nodes activate their neighboring nodes with a given probability, and once active

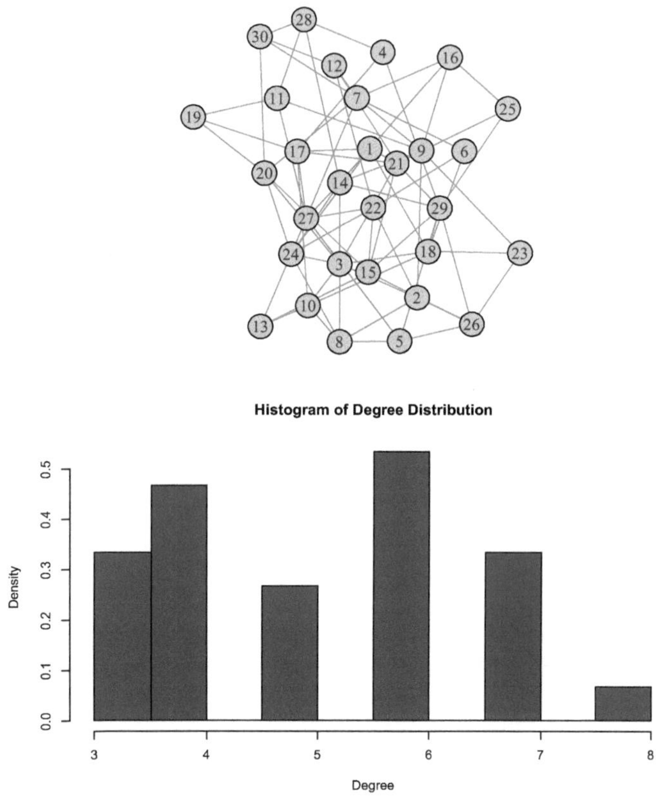

Fig. 8.4 A sample network generated from the Erdos-Renyi random graph model [11, 12] with $N = 30$ nodes and connection probability $p = 0.2$. The left panel shows the network structure. Right panel shows the degree distribution as a histogram of node degrees. The degree distribution is approximately Poisson, and the mean degree is 5.13

a node remains active. In the spiking process, each node fires an action potential ("spike") at random times. (For readers who have taken a course on stochastic processes, the random times are selected according to a Poisson process. A Poisson process is a simple continuous-time stochastic process, also known as a counting process, that counts events during a given time interval [26].) After an action potential spike occurs, a node excites its neighboring nodes (often causing those nodes to fire action potentials) and then its activity level decays exponentially to a baseline low level of activity until it fires again. Thus, in the spiking model, nodes change their activity level (increasing or decreasing their firing rates) depending on the interactions it has with neighboring nodes. In contrast, once a node becomes active in the activation process, it can never change its activity.

The main question we investigate is to what extent the graph structure is reflected in the dynamics of the process on the graph. Our findings suggest that memoryless processes, such as the activation process described above, are more likely to reflect

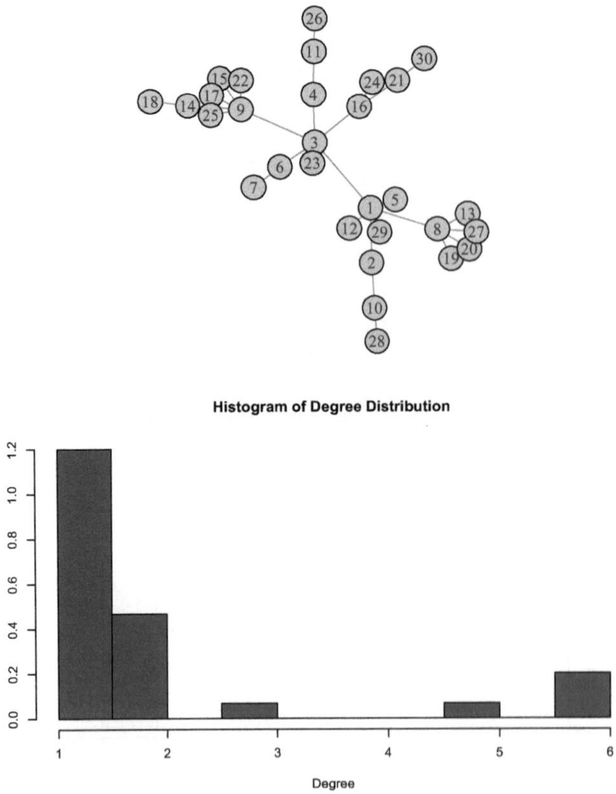

Fig. 8.5 A sample scale-free network generated from the preferential attachment model [3] with $N = 30$ nodes. The left panel shows the network structure, note that most nodes only have 1 neighbor but a few nodes have many neighbors. Right panel shows the degree distribution as a histogram of node degrees. The degree distribution approximately follows a power law, and the mean degree is 1.93

the degree distribution. We refer to these processes as memoryless because the nodes show neither potentiation nor depression. That is, in response to being stimulated to fire frequently, they neither increase their response nor do they adapt to the stimulus and show a dampened response. We find that processes with memory, such as the spiking process described above in which nodes change their firing rate in response to inputs, can robustly produce power law and other heavy-tailed distributions of activity regardless of degree distribution.

To connect the work above to a particular biological system, consider sleep-wake regulation. Sleep-wake behavior in mammals is governed by interacting networks of neurons in the brain that can be well-modeled as stochastic processes on random graphs. Recent studies of sleep-wake dynamics have shown that, in several mammalian species, lengths of periods of wakefulness (known as wake bouts) follow a power law distribution, while sleep bouts follow an exponential distribution; see, for instance, [24, 28]. Joshi and Best looked at a model of three interacting neuronal populations (or nodes in a 3-node graph) that mimics the dynamics of sleep-wake cycling in developing rats: It produces short, exponentially distributed wake and sleep bouts before transitioning to consolidated wake and sleep episodes with power law distributed wake bouts [18].

The spiking model we considered in [35] can be viewed as expanding the wake-active population node from the model by Joshi and Best into a network of individual spiking neurons where the activity of each node is described by a stochastic differential equation. Currently, we are extending our model to link network structure and stochastic dynamics to neural activity patterns involved in sleep-wake regulation [38]. Focusing on the study of two interacting networks, such as the dynamics between sleep-active and wake-active neural networks, we have a spiking model that assumes excitatory connections within each network and inhibitory connections between the two networks. See Fig. 8.6 for an illustration of two interacting networks where each network is generated independently from the Erdos-Renyi random graph model. For the case where both networks have degree distributions with low mean degree, the model robustly produces wake bout durations distributed as a power law followed by an exponential tail. The size of the power law region and the variability of the power law exponent across different realizations of the model depends on the network structure.

In empirical studies, the distribution of wake bout durations is highly dependent on the age of the mammal so a related question is: What developmental changes are influencing the distribution of wake bouts? We plan to address this question in future work. It is important to keep in mind that research in mathematical biology includes an iterative development of a model, comparison with known aspects of the biological system being studied, and ultimately using the model to generate new questions and observations.

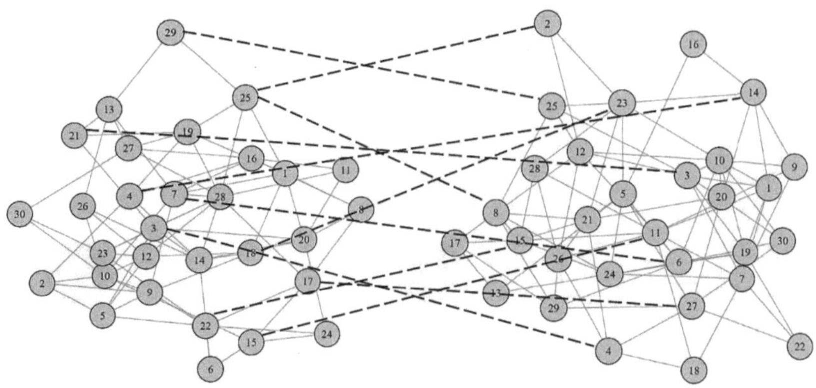

Fig. 8.6 An illustration of two interacting networks, each randomly generated from the Erdos-Renyi random graph model with $N = 30$ and $p = 0.2$. The solid edges present within each of the two networks represent excitatory connections between neurons. The dashed edges indicate inhibitory connections between the networks

For more information on network structure and dynamics of biological systems, I have written a detailed book chapter on this topic with an emphasis on open and accessible problems for undergraduate research geared for students and their advisors [32]. My chapter also includes substantial background material needed to study this area.

8.3.2.2 Identifying Important Sources of Noise in Neural Systems

Another topic of growing importance in mathematical neuroscience is the inclusion of stochastic effects within dynamical neural models. Nerve cells produce electrical impulses (action potential "spikes") through the coordinated opening and closing of ion channels located in the membrane of each nerve cell. A major source of noise in neural systems is the random gating of ion channels, and such systems are typically represented as a stochastic (Markov) process on a graph. Markov processes with variable transition rates (i.e., rates that depend on the voltage of the cell) capture the stochasticity of spike generation, but at the cost of complex, time-consuming simulations. Discrete-state, continuous-time Markov processes occur throughout cell biology and neuroscience representing the random dynamics of processes transitioning among multiple locations or states. Complexity reduction for such models aims to capture the essential dynamics and stochastic properties via a simpler representation, with minimal loss of accuracy. Many researchers have worked on reducing complexity by eliminating variables or replacing the graph with a simpler one. Figure 8.7 gives an example of replacing a graph (left) with a simpler graph (right) by grouping all white nodes together into one. However, these approaches can break the statistical independence between unconnected nodes (in other words, the reduced model no longer satisfies the Markov property), making such models harder to analyze.

The stochastic shielding approximation is a new approach to efficient, accurate simulation of Markov chain models [29]. This approximation works by neglecting

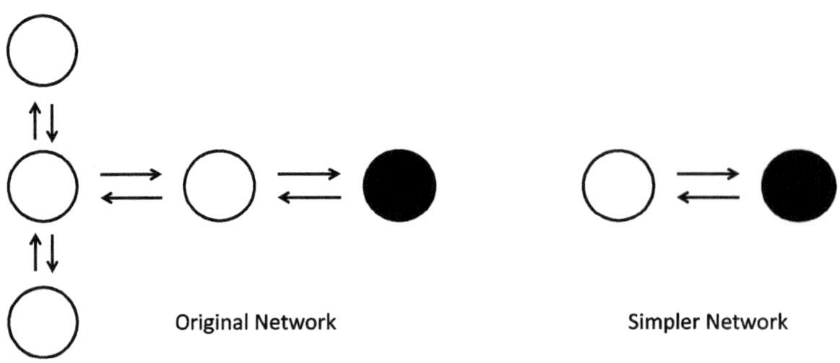

Fig. 8.7 Left panel: original network of 5 nodes. Right panel: simpler network of 2 nodes made by grouping the white nodes from the left panel together into one white node

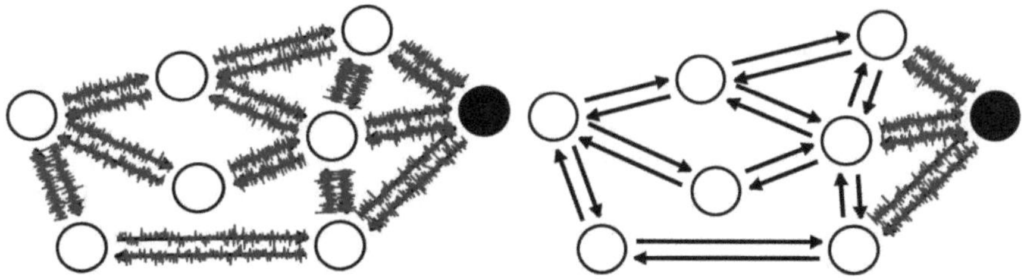

Full Process:
Include mean fluxes and all
sources of fluctuations.

Approximate Process:
Include mean fluxes and only
selected sources of fluctuations.

Fig. 8.8 An illustration of the stochastic shielding (SS) approximation. The left panel represents the full stochastic process on the network (24-dimensional). The right panel represents the approximate stochastic process (six-dimensional) on the network according to the SS rule: Suppress the noise along edges connecting indistinguishable nodes (white-white node connections in this figure). Note that the network (graph) is the same in both cases. (From [36], licensed under CC-BY 4.0.)

a carefully chosen subset of noise sources within the Markov process and leads to significantly faster simulations with practically no observable error. Inspired by this work and in collaboration with Peter Thomas (CWRU), I conducted a thorough mathematical analysis of the stochastic shielding approximation [34]. This method is based on replacing a high-dimensional stochastic process defined on a graph with a lower-dimensional process on the same graph, rather than replacing a complex network with a simpler one. See Fig. 8.8 for an illustration. The network has 8 nodes and 24 edges so a stochastic process on this network will be 24-dimensional, one dimension for each edge in the network. The wavy line on an edge represents a transition modeled by a one-dimensional stochastic process. Suppose one node, the one colored black, is distinguishable from the other nodes. Then we can replace the 24-dimensional stochastic process on the left with a six-dimensional stochastic process on the right, where the other 18 transitions are deterministic (meaning that the noise is suppressed and a mean rate of transition is used). The deterministic transitions in the network on the right do not have wavy lines, and those edges connect so-called *indistinguishable nodes* , represented by two white nodes. The stochastic shielding method is based on the following rule: Suppress the noise along the edges connecting indistinguishable nodes.

We showed that this form of model reduction can be represented as a mapping from the original process to an approximate process defined on a significantly smaller sample space. Our analysis results in a new, quantitative measure of the importance of individual edges within a Markov process on a graph. We called this the *edge importance measure*; it extends the stochastic shielding method by quantifying the effect of suppressing noise along each edge separately [34]. Moreover, as we show below, the edge importance measure allows us to rank the edges in the graph in order of importance relative to a specific type of variance.

We refer to the black node in Fig. 8.8 as the *observable state*. Biologically, this could be the conducting state in an ion channel model where the flow of ions across the cell membrane can be measured (observed), for instance. We are interested in computing the variance in node occupancy for this observable state under a special condition called stationarity; this type of variance is called the *stationary variance* and relies on the stationary distribution of the Markov process. The term "stationary" refers to the long-term behavior of the system. Informally, a stationary distribution of a Markov chain is a probability distribution that remains unchanged as the Markov chain moves forward in time. Formally, the stationary distribution of a Markov chain with transition matrix P is a vector, π, such that $\pi P = \pi$. This means that, in the long run, no matter in what state the process starts, the proportion of time the process spends in state j is approximately π_j for all states $j = 1, 2, \ldots, n$ in the state space. We also refer to $\pi = [\pi_1 \pi_2 \ldots \pi_n]$ as the stationary (or steady state or equilibrium) node occupancy probability vector.

A key contribution of the work in [34] is showing that the stationary variance of the observable state can be decomposed into a sum of fluctuations attributable to each pair of directed edges in the graph. To illustrate this stationary variance decomposition, I will use the simple 3-state chain shown in Fig. 8.9. For simplicity, I will assume the transition rates are $\alpha_k = 1$ for all $k \in \{1, 2, 3, 4\}$. A fixed number of ion channels $N \geq 1$ (in simulations we used $N = 100$) move between the three states, where at each time step, each state is occupied by some portion of the ion channels. State 3 (black node) is the observable state and we are interested in the stationary variance R of this state, defined in Eqs. 8.14–8.15. Here, the stationary variance is the variance of the number of ion channels in the observable state over time.

Before stating the definition of the stationary variance R of the observable state, we first need to define several quantities related to the graph in Fig. 8.9. The adjacency matrix A of this graph is given below along with matrix D, the diagonal matrix of node out-degrees:

$$A = \begin{pmatrix} 0 & 1 & 0 \\ 1 & 0 & 1 \\ 0 & 1 & 0 \end{pmatrix} \quad \text{and} \quad D = \begin{pmatrix} 1 & 0 & 0 \\ 0 & 2 & 0 \\ 0 & 0 & 1 \end{pmatrix}. \tag{8.9}$$

From these two matrices, we can define the graph Laplacian $L = (A - D)^T$, which plays a key role in the decomposition of the stationary variance of the observable state.

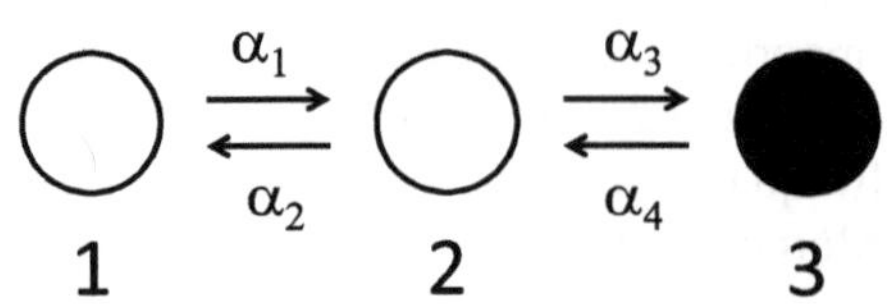

Fig. 8.9 Three-state Markov chain with transition rates α_k for $k \in \{1, 2, 3, 4\}$. State 3 (black node) is the observable state

$$L = \begin{pmatrix} -1 & 1 & 0 \\ 1 & -2 & 1 \\ 0 & 1 & -1 \end{pmatrix} \tag{8.10}$$

Recall that we explicitly keep track of each edge k; edge k represents a transition from source node $i(k)$ to destination node $j(k)$. We do this to compute various quantities corresponding to each edge. Now we can define the noise matrix B for the graph in Fig. 8.9, where the rows of B correspond to the three states and the columns of B correspond to the four numbered edges. More specifically, each column of B corresponds to a stochastic transition along each edge. For instance, column 1 represents a weighted transition from state 1 to state 2, column 2 represents a weighted transition from state 2 to state 1, and so on. The weight will be defined after the definition of B. If a column is all zeros, that indicates a deterministic transition; this is equivalent to suppressing the noise along that edge and only looking at the average behavior or mean flux (see Fig. 8.8). Thus, we have the noise matrix:

$$B = \begin{pmatrix} -\sqrt{J_1} & \sqrt{J_2} & 0 & 0 \\ \sqrt{J_1} & -\sqrt{J_2} & -\sqrt{J_3} & \sqrt{J_4} \\ 0 & 0 & \sqrt{J_3} & -\sqrt{J_4} \end{pmatrix} \tag{8.11}$$

where $J_k = N\pi_{i(k)}$ is the stationary mean flux[2] along the kth edge, N is the total number of ion channels in the population, and the stationary node occupancy probability[3] $\pi = [\pi_1, \pi_2, \pi_3]$ is uniform, or $\pi = [1/3, 1/3, 1/3]$.

The stationary variance definition below also uses the eigenvalues λ_i and eigenvectors v_i of the graph Laplacian matrix L and two other bookkeeping variables: M and ζ_k. M is a three-dimensional vector labeling the nodes as observable (1) or unobservable (0) states, or black versus white nodes in the figures above. For the 3-state process,

$$M = \begin{pmatrix} 0 \\ 0 \\ 1 \end{pmatrix} \tag{8.12}$$

since state 3 is the only observable state. The vector ζ_k describes the transition for edge k in the graph for $k \in \{1, 2, 3, 4\}$, and weighted versions of these four vectors are the columns of matrix B above (weighted by the square root of the stationary flux). For the 3-state process,

[2] Recall that the term "stationary" refers to the long-term behavior of the system and is formalized with the stationary distribution. The stationary mean flux along edge k is the average number of state transitions along edge k over time. For instance, if $N = 100$ and $\pi = [1/3, 1/3, 1/3]$, then $J_k = N\pi_{i(k)} = 100/3$ for each edge k.

[3] The stationary occupancy probability of node i is the long-term probability of a single ion channel being in state i.

$$\zeta_1 = \begin{pmatrix} -1 \\ 1 \\ 0 \end{pmatrix} = -\zeta_2 \quad \text{and} \quad \zeta_3 = \begin{pmatrix} 0 \\ -1 \\ 1 \end{pmatrix} = -\zeta_4. \tag{8.13}$$

Finally, we can describe the stationary variance R of the observable state 3. It is given by the following decomposition:

$$R = \sum_{k=1}^{4} R_k \tag{8.14}$$

where

$$R_k = J_k \sum_{i=2}^{4} \sum_{j=2}^{4} \left(\frac{-1}{\lambda_i + \lambda_j} \right) \left(M^T v_i \right) \left(v_i^T \zeta_k \right) \left(\zeta_k^T v_j \right) \left(v_j^T M \right). \tag{8.15}$$

If we compute R_k for $k \in \{1, 2, 3, 4\}$, we see that $R_1 = R_2 = 0.0417$ and $R_3 = R_4 = 0.2917$. The total variance, or sum of these four quantities, is $R = 0.6668$. Hence, it turns out that edges 3 and 4 contribute the most to the stationary variance of state 3, or 7/8 of the total variance R, since

$$\frac{R_3 + R_4}{R} = 0.825 = \frac{7}{8}. \tag{8.16}$$

Edges 1 and 2 only contribute 1/8 of the total variance since

$$\frac{R_1 + R_2}{R} = 0.175 = \frac{1}{8}. \tag{8.17}$$

Therefore, the noise from transitions 1 and 2 is minimal and can be suppressed to accurately approximate the fully stochastic 3-state process corresponding to Fig. 8.9. This is exactly what the stochastic shielding method says to do to make the most accurate approximation for the 3-state process that preserves the dynamics of the observable state 3.

We described the stochastic shielding method for a simplistic toy model, but this method can be used for much more complicated systems. The edge importance measure allows us to rank the edges in a network according to which ones are the most important (with respect to the stationary variance described above) so that we can choose how much noise to include in simulations of the model in terms of maximizing both speed and accuracy. This measure not only confirms the optimality of the stochastic shielding approximation, but when applied to problems in neuroscience, it also sheds new light on the contributions of different ion channel transitions to the variability of neural systems.

Much of my research has involved applying the stochastic shielding method to different real-world processes, such as sodium and potassium ion channels, which are important for understanding neuroscience. I have also investigated how well the stochastic shielding method performs for more complicated situations, such as the nicotinic acetylcholine receptor, an ion channel that exhibits "bursty" behavior [25, 36]. This means that the ion channel exhibits bouts of repeated channel opening and closing interspersed by long periods of channel closure, so one observes bursts of activity instead of constant activity. Mathematically, we consider processes with nonuniform stationary probabilities and multiple timescales, as an extension to our previous work. In [36], we establish that the edge importance measure remains a valid tool for analysis for arbitrary networks regardless of multiple timescales.

8.3.3 Fitting Models to Data

As experimentalists collect more and more data, it becomes increasingly important to understand the extent to which we can use modeling frameworks, like those described above, as statistical models. For example, given simulated data from one of these models, how well can we estimate the parameters used to simulate the data?

Projects that focus on fitting models to data are varied and not necessarily all related to biology. I have worked on projects that range from fitting the frequency distribution of words from a very large data set of English texts to a power law distribution [2] to modeling the activity of gonadotropin releasing-hormone neurons in rats [13] to using likelihood methods to investigate the diversification of zinc-finger binding domains in the human genome [33]. Most recently, I used clustering methods to classify the contributions of olfactory receptor neuron activity to navigational behavior in Drosophila larvae [6], and I fit logistic growth curves to amphibian pathogen data from varied geographical locations [39]. I also worked with my Master's student to develop a new method for analyzing psychomotor vigilance task (PVT) data used in sleep research [1].

The projects described above all have a common theme of fitting mathematical and/or statistical models to real data. As a mathematical biologist, working with real data is always very interesting and helps to inform new mathematical questions to investigate. For instance, the power law paper [2] has led me to study several new mathematical questions: How does the choice of probability distribution influence the estimated power law exponent for a given data set? How do the estimates compare when using truncated versus untruncated forms of these different probability distributions? This work is currently in preparation [37]. A related mathematical question is how to identify power law regimes within a given data set, or subintervals of the data that follow a power law distribution? A recent collaborative paper of mine [23] yields a new method by adaptively penalizing the Kolmogorov-Smirnov (KS) method to identify power law distributed intervals in data. The classical KS method optimizes the Kolmogorov-Smirnov distance between presumed power law fits and the random sample, over proposed tail regions, to estimate the lower bound of the tail region. Our work extends the KS method to identify power law regimes bounded on both ends of the distribution.

8.4 Future Work

There are many ways I plan to extend the work I have discussed here. First, there are several natural extensions to my work on edge importance measures for biological network models. We obtained mathematically rigorous results in [34, 36] by exploiting the stationarity of the stochastic ion channel process, corresponding to a common experimental condition known as "voltage clamp" of an electrically active nerve cell. A major challenge that I plan to address is to extend our analysis to the *nonstationary* case, corresponding in experimental terms to "current clamp." Under this condition, the cell voltage is allowed to vary over a significant range as the cell "spikes" or produces action potentials. The transition rates of the stochastic process depend on the cell voltage, and hence, the ion channel process becomes strongly nonstationary. Nevertheless, it has been shown empirically that the same stochastic shielding method works equally well under stationary conditions (for which we have rigorous results [34, 36]) and nonstationary conditions [29]. In order to address the latter case, we need to develop new mathematical ideas.

I also plan to extend the sleep-wake model described above and investigate problems pertaining to the genetic regulation of sleep. In particular, interesting extensions include studying how sleep-wake dynamics change during neural development alongside the developing neuronal network structure and also how differences in gene regulation and function may influence sleep-wake patterns. I am currently working on statistical questions surrounding the emergence of power law distributions in biological data, including sleep data.

This approach of delving deep into the biology to understand what questions need to be answered is fundamental to my research in mathematical biology. Biology certainly can and does lead to new mathematics. The primary means by which it does so is by raising questions that mathematicians had not thought to ask. Once the mathematical question arises, we often find that the mathematical theory necessary to address the question has not yet been developed.

References

1. Armstrong, J.M.: Developing a general method for analyzing psychomotor vigilance task (PVT) data: modeling sleep inertia in children. Master's thesis, University of Nevada, Reno (2018)
2. Arnold, R., Schmidt, D.: Changes in the coefficients of Zipf's law for English corpora of different contexts. Nevada State Undergrad. Res. J. **3**(1), 14–25 (2017)
3. Barabási, A.L., Albert, R.: Emergence of scaling in random networks. Science **286**(5439), 509–512 (1999)
4. Behe, M.: The Edge of Evolution: The Search for the Limits of Darwinism. Simon and Schuster, New York (2007)
5. Calabrese, P., Mecklin, J.P., Järvinen, H.J., Aaltonen, L.A., Tavaré, S., Shibata, D.: Numbers of mutations to different types of colorectal cancer. BMC Cancer **5**, 1–7 (2005)

6. Clark, D.A., Odell, S.R., Armstrong, J.M., Turcotte, M., Kohler, D., Mathis, A., Schmidt, D.R., Mathew, D.: Behavior responses to chemical and optogenetic stimuli in Drosophila larvae. Front. Behav. Neurosci. **12**, 324 (2018)
7. Durrett, R., Schmidt, D.: Waiting for regulatory sequences to appear. Ann. Appl. Probab. **17**(1), 1–32 (2007)
8. Durrett, R., Schmidt, D.: Waiting for two mutations: with applications to regulatory sequence evolution and the limits of darwinian evolution. Genetics **180**(3), 1501–1509 (2008)
9. Durrett, R., Schmidt, D.: Reply to Michael Behe. Genetics **181**(2), 821–822 (2009)
10. Durrett, R., Schmidt, D., Schweinsberg, J.: A waiting time problem arising from the study of multi-stage carcinogenesis. Ann. Appl. Probab. **19**, 676–718 (2009)
11. Erdös, P., Rényi, A.: On random graphs. Publ. Math. **6**, 290–297 (1959)
12. Erdös, P., Rényi, A.: On the evolution of random graphs. Publ. Math. Inst. Hung. Acad. Sci. **5**(1), 17–60 (1960)
13. Gay, V.L., Hemond, P.J., Schmidt, D., O'Boyle, M.P., Hemond, Z., Best, J., O'Farrell, L., Suter, K.J.: Hormone secretion in transgenic rats and electrophysiological activity in their gonadotropin releasing-hormone (GnRH) neurons. Am. J. Physiol. Endocrinol. Metabol. **303**(2), E243–E252 (2012)
14. Gunter, C., Dhand, R.: The chimpanzee genome. Nature **437**(7055), 47 (2005)
15. Hartl, D.L., Clark, A.G.: Principles of Population Genetics, 4th edn. Sinauer Associates, Sunderland (2006)
16. Hurtado, P.J.: Building new models: rethinking and revising ODE model assumptions. In: Highlander, H.C., Capaldi, A., Eaton, C.D. (eds.) Foundations for Undergraduate Research in Mathematics: An Introduction to Undergraduate Research in Computational and Mathematical Biology: From Birdsongs to Viscosities, pp. 1–86. Springer, Berlin (2020)
17. Hurtado, P.J., Kirosingh, A.S.: Generalizations of the 'linear chain trick': incorporating more flexible dwell time distributions into mean field ode models. J. Math. Biol. **79**(5), 1831–1883 (2019)
18. Joshi, B.S.: A doubly stochastic Poisson process model for wake-sleep cycling. Ph.D. thesis, The Ohio State University (2009)
19. Kumar, S., Filipski, A., Swarna, V., Walker, A., Hedges, S.B.: Placing confidence limits on the molecular age of the human–chimpanzee divergence. Proc. Natl. Acad. Sci. **102**(52), 18842–18847 (2005)
20. Luebeck, E.G., Moolgavkar, S.H.: Multistage carcinogenesis and the incidence of colorectal cancer. Proc. Natl. Acad. Sci. **99**(23), 15095–15100 (2002)
21. Moran, P.A.P.: Random processes in genetics. In: Mathematical Proceedings of the Cambridge Philosophical Society, vol. 54, pp. 60–71. Cambridge University Press, Cambridge (1958)
22. Nowak, M.A.: Evolutionary Dynamics: Exploring the Equations of Life. Belknap Press, Cambridge (2006)
23. Olmez, F., Kramer, P.R., Fricks, J., Schmidt, D.R., Best, J.: Penalized KS method to fit data sets with power law distribution over a bounded subinterval. J. Stat. Comput. Simul. **91**(8), 1524–1563 (2021)
24. Patel, M., Rangan, A.: Role of the locus coeruleus in the emergence of power law wake bouts in a model of the brainstem sleep-wake system through early infancy. J. Theor. Biol. **426**, 82–95 (2017)
25. Purves, D., Augustine, G., Fitzpatrick, D., Hall, W., LaMantia, A., McNamara, J., White, L.: Neuroscience, 6th edn. Sinauer Associates, Sunderland (2017)
26. Ross, S.M.: Introduction to Probability Models, 11th edn. Academic Press, Cambridge (2014)
27. Rudin, W.: Real and Complex Analysis, 3rd edn. McGraw-Hill Education, New York (1987)
28. Sandberg, A.J., Schmidt, D.R.: An integrate-and-fire mathematical model of sleep-wake neuronal networks in the developing mammal. PLoS ONE **19**(10), e0307851 (2024)
29. Schmandt, N.T., Galán, R.F.: Stochastic-shielding approximation of Markov chains and its application to efficiently simulate random ion-channel gating. Phys. Rev. Lett. **109**(11), 118101 (2012)

30. Schmidt, D.R.: A mathematical look at DNA regulatory sequence evolution. Ph.D. thesis, Cornell University (2007)
31. Schmidt, D.R.: Balancing an academic career in mathematics with motherhood and life's other passions. J. Humanistic Math. **8**(2), 281–291 (2018)
32. Schmidt, D.R.: Network structure and dynamics of biological systems. In: Highlander, H.C., Capaldi, A., Eaton, C.D. (eds.) Foundations for Undergraduate Research in Mathematics: An Introduction to Undergraduate Research in Computational and Mathematical Biology, pp. 299–351. Springer, Berlin (2020)
33. Schmidt, D., Durrett, R.: Adaptive evolution drives the diversification of zinc-finger binding domains. Mol. Biol. Evol. **21**(12), 2326–2339 (2004)
34. Schmidt, D.R., Thomas, P.J.: Measuring edge importance: a quantitative analysis of the stochastic shielding approximation for random processes on graphs. J. Math. Neurosci. **4**(6), 1–52 (2014)
35. Schmidt, D., Best, J., Blumberg, M.S.: Random graph and stochastic process contributions to network dynamics. In: Discrete and Continuous Dynamical Systems, vol. Suppl. 2, pp. 1279–1288 (2011)
36. Schmidt, D.R., Galán, R.F., Thomas, P.J.: Stochastic shielding and edge importance for Markov chains with timescale separation. PLoS Comput. Biol. **14**(6), e1006206 (2018)
37. Schmidt, D.R., Hurtado, P.J., Forister, M.L.: Not all alphas are created equal: fitting power law distributions to empirical data. In preparation
38. Schmidt, D.R., Olmez, F., Kramer, P.R., Best, J.: Bout time distributions in a symmetric sleep-wake model. In preparation
39. Sheets, C.N., Schmidt, D.R., Hurtado, P.J., Byrne, A.Q., Rosenblum, E.B., Richards-Zawacki, C.L., Voyles, J.: Thermal performance curves of multiple isolates of Batrachochytrium dendrobatidis, a lethal pathogen of amphibians. Front. Vet. Sci. **8**, 687084 (2021)
40. Wootton, J.C., Feng, X., Ferdig, M.T., Cooper, R.A., Mu, J., Baruch, D.I., Magill, A.J., Su, X.z.: Genetic diversity and chloroquine selective sweeps in plasmodium falciparum. Nature **418**(6895), 320–323 (2002)

Chapter 9
Finding Your Path

Abra Brisbin, Karen Lange, Erin McNicholas, and Emilie Purvine

Abstract We hope that this volume has given you a taste of the wide variety of topics mathematicians study, the varied paths they take, and how you might chart your own path through graduate school and beyond. In this last chapter, we, the editors, share our career paths and particularly how we came to our graduate research specialties and how our research interests have evolved over time. Karen's biography highlights the importance of trying things out along the way. In Emilie's biography, you will read how her path (or "meander" as she calls it) led to career lessons about the importance of communication and the ability to reinvent yourself. Erin shows by example that finding something *you* are passionate about leads to a fulfilling career. And finally, Abra provides a detailed walkthrough of her journey from undergraduate student to department chair together with advice from that lived experience.

Keywords Career paths · Career advice · Graduate school · Mathematical community · Choosing a research subfield · Carleton Summer Math Program · Exploring interests · Mentorship · Dealing with setbacks · Applying for jobs · Work-life balance · Taking risks · Research experiences · Self-advocacy

---

A. Brisbin
University of Wisconsin, Eau Claire, WI, USA
e-mail: brisbia@uwec.edu

K. Lange (✉)
Wellesley College, Wellesley, MA, USA
e-mail: klange2@wellesley.edu

E. McNicholas
Willamette University, Salem, OR, USA
e-mail: emcnicho@willamette.edu

E. Purvine
Pacific Northwest National Laboratory, Seattle, WA, USA
e-mail: emilie.purvine@pnnl.gov

A. Brisbin et al. (eds.), *Research Connections*, Association for Women in Mathematics Series 36, https://doi.org/10.1007/978-3-031-81217-0_9

Ours and the authors' autobiographies highlight some general advice for choosing a research direction that we summarize here:

- **Explore widely** in both your undergraduate and early graduate school years. Such exploration allows you to get a sense of different areas of math and gives you a solid foundation of tools to draw from as your interests evolve. Specifically:
  - Take a variety of classes.
  - Attend colloquia and seminars.
  - Attend conferences and summer schools.
  - Look into multiple career options—internships being one great way to do this!

  Keep in mind the breadth and intended audience of different classes, colloquia, and conferences, as some are more specialized and narrowly focused than others. Early on in your career look for experiences that cover a wider range of topics (e.g., the Joint Mathematics Meetings) or are aimed more specifically at people in your career stage (e.g., conferences aimed at undergraduates or graduate students).
- **Reach out to learn more about areas you are interested in.**
  - *Speak with professors who work in the area.* Ask them for accessible introductions to their work (whether a text or a survey paper) and chat with them about where their work is heading and possible opportunities for undergraduate or graduate students. Find out what classes or skills they recommend for a student interested in this area. Ask if they are willing to meet occasionally to discuss a paper related to their work, or begin attending more focused seminars they run on the topic. Larger research groups often have a seminar focused on their area in which graduate students read and present papers in the field. If your advisor is not part of a group or does not have many graduate students, this kind of work is likely to be done one on one. Such informal experiences will help you understand what working with a particular faculty member will be like.
  - *There is no need to rush to decide on an area or advisor.* As long as you remain on track with your program, take advantage of the freedom and structure of your early graduate school years to explore areas and determine whether you and your potential advisor work well together.
- **Find and nurture your mathematical community.**
  Keep in mind that you are not just choosing a research topic for your dissertation; you are also choosing a mathematical community (both within your program and outside of it). These communities vary by size and working culture (e.g., how collaborative is the field, how supportive of early career mathematicians, etc.).
  - *Cultivate professional relationships with people at all career stages,* from fellow graduate students to senior members of your field. Conferences aimed at graduate students in your field are a great place to get to know your peers,

many of whom will be your colleagues throughout your career. You can get to know faculty outside of your institution (and their work) by reading their papers and approaching them at conferences and summer programs, or when they come to visit your department. Having more contacts and mentors provides you a wider spectrum of advice and perspectives and opens up opportunities for collaboration, speaking invitations, and positions.

- **You can always change your path.**
  Changing research or career direction is not always easy, but people make these changes all the time. If some aspect of your path is not a good fit, brainstorm with your mentors about possible changes and make (calculated) adjustments.
- **Be true to yourself.**
  In the course of your education and career, you will encounter many people who followed a similar path to a similar destination. Sometimes it can seem that there is only one path or only one destination, when in fact there are a multitude of ways to become an academic and a multitude of careers that incorporate math skills in a meaningful way outside of academia. Do not let the values and aspirations of others dictate what your goals and values should be. Often, well-intentioned mentors (ourselves included!) will offer advice that reflects their own experiences, but may not be the right advice for you. Do not be afraid to chart your own course or let others know when your goals and needs are different. Remember that you deserve to be part of a community that values you for who you are. We all bring different skills and perspectives, and it is the combination of all these approaches and insights that propels mathematics forward. If you feel like you have to present as someone you are not, or that your skills and talents are not valued, that is a problem with the environment you are in, not a problem with you. You belong in the community of mathematicians and you are entitled to leave current advisors, departments, and positions for other advisors, departments, and positions that will value you.

As you read through the editors' biographies below and those of all of the authors in this book, you may see parts of your own journey-in-progress. The message we hope you take away is that we all get to where we are going eventually. While there is much you can learn from other people's experiences, your experience will be unique. That is a good thing! By bringing your own perspectives and talents to mathematics, you make our mathematical community stronger.

9.1 Karen Lange: My Path to Computability Theory

I work in computability theory, a branch of mathematical logic that answers "What does it mean to compute?" and uses related ideas to measure the complexity of mathematical problems. But I definitely did not follow any algorithm to find my way to this field, though I am glad I landed here.

Before college, I enjoyed mathematics, but I loved many things, from theater to bowling to history. So, I was pleasantly surprised when I ended up majoring in math after repeatedly taking more math classes each semester. My professors at Swarthmore College encouraged me to apply to Carleton's Summer Math Program for Women (SMP) during my freshman year. Get paid to learn dynamical systems and game theory over the summer on Carleton's idyllic campus in Minnesota?! Sure! At SMP, I met others on their own math journeys as well as supportive faculty who opened my eyes to even more opportunities. Over the next two summers, I got to try out math research at two Research Experiences for Undergraduates (REUs), one in matrix theory at William and Mary and one in cluster analysis at the DIMACS institute at Rutgers University. I applied to programs that explored topics that appealed to me, but I began to realize these programs also provided opportunities to live in different parts of the country and even the world. For example, the DIMACS REU had an international exchange component with Charles University in the Czech Republic, which sounded amazing especially since I had planned to study in neighboring Hungary at the Budapest Semesters in Mathematics program during the fall of my senior year. As part of the DIMACS REU, I ended up traveling to a math conference at Charles University in Prague, before moving on to Hungary for the semester. These experiences opened my eyes to the wider math community and to what being an academic mathematician entailed.

At Swarthmore, I kept taking math classes and served as a grader for the department. To build a skill set of widely applicable tools, I figured I should learn some computer science and statistics. After learning some programming, courses in algorithms and the theory of computation appealed to my mathematical sensibilities (and completed my computer science minor). One question on the theory of computation final asked about the feasibility of creating code debugging software that could catch all possible programming errors one could potentially make. The novel (to me) question provoked a personal epiphany about the limitations of computation imposed by the *halting problem*, a key course concept that is at the heart of my research discipline. This experience gave me a deeper appreciation of the creativity that goes into excellent teaching. My professors' jobs began to strike me as amazing; they guided us students to learn beautiful mathematics while they themselves had the freedom to mold their own mathematical journey through research, often in collaboration with colleagues across the globe. I was thrilled to learn that most PhD programs in math were fully funded; programs typically cover tuition and provide a stipend to cover modest living expenses. Somewhere in my junior year I began to plan to attend graduate school in math, so that I could pursue these newfound academic passions.

Like most students heading into graduate school, I was not sure what area I wanted to pursue in mathematics. Hence, I looked for midsize or larger programs that required studying the major subfields of mathematics. Given how much I enjoyed the theory of computation, I did look for programs that had faculty working in related areas, such as discrete mathematics and logic (even though I had not taken an undergraduate course in logic!). I also looked for programs that would provide me substantial pedagogical opportunities, since I envisioned teaching and working closely with students as central aspects to my dream job. I was lucky to get into a few programs that fit these constraints and decided on University of Chicago (U of C), in part for its geographic proximity to Milwaukee, where my family was.

The first year program at U of C consisted of a predetermined set of three courses each in analysis, algebra, and geometry/topology. While these courses gave me a firm grounding in their topic, I still felt unsure about where I wanted to head in terms of research. Though logic was not a part of the first year curriculum, I approached Robert Soare about exploring computability theory in the spring. I was startled when he immediately began to plot my path into logic. A senior graduate student would be speaking in the logic seminar the following week, and he stated that I should speak before her to introduce a tool she would be applying in her work. He pulled out an extra copy of the text he had authored and began explaining the portion I should cover. Having no previous background in the area, I dove into this new world and managed to present the assigned result before the senior graduate student's lecture. From that point on, I was a member of the computability group.

Robert Soare has a strong point of view about how graduate students should be trained. He ensured that his graduate students had substantial interaction with the many visitors he arranged and that there was funding to attend conferences and give talks. These opportunities grew my professional network and introduced me to the many avenues people were pursuing in the field. Denis Hirschfeldt, whose research interests overlapped substantially with Soare's, was my second advisor. Having two advisors with complementary research interests and approaches provided me double the ideas, advice, and encouragement. Moreover, there were enough graduate students in the logic group to create a learning community and sense of camaraderie. I found the wider computability community welcoming, collaborative, and supportive of junior people.

I had been drawn to U of C in part because the program provided graduate students substantial opportunities to teach. After my first year at U of C, I spent a year serving as a teaching assistant before regularly teaching as a stand-alone instructor. (I TA'ed for the undergraduate logic course, a great way to learn the material!) But there were many other pedagogical opportunities, too. I began advising undergraduate independent studies as part of the Directed Reading Program and later helped to administer the program itself. Some faculty worked with teachers in the Chicago Public Schools. Through their programs, I taught professional development courses for teachers and got to do math with students in local elementary schools. Finally, the U of C's Center for Teaching gave me the chance to discuss teaching with graduate students from across the University. I did not start out knowing about all these possibilities, and often taking part in one led to another. My route through these pedagogical experiences was unplanned, but taken together, they organically strengthened my teaching abilities and widened my networks and professional horizons.

Even though I found these activities fulfilling, I was still on the fence about pursuing an academic career as I neared the end of my doctoral studies. I wondered what nonacademic paths could offer, since I really had not explored them. Furthermore, at U of C, I had met and (a few years later) married Michael, a fellow classmate studying number theory. Michael was also contemplating an academic career, and the challenge of navigating a "two body problem" (i.e., landing two permanent academic jobs reasonably near to one another) seemed daunting. Through the career services center at U of C, I learned about industry opportunities and took on an internship at a local software firm the summer before I earned my doctorate. While I enjoyed solving concrete, relevant problems on a daily basis during my internship, I also missed teaching and the freedom of academic life. Exploring my nonacademic options helped me gain clarity about my next steps. Ultimately I decided to pursue an academic career.

I applied to both postdoctoral and teaching positions during my last year of graduate school. I also "threw my hat in the ring" for the National Science Foundation (NSF) Postdoctoral Fellowship (again on the wise counsel of my advisors). I chose to apply to work with Julia Knight and Peter Cholak, two members of the University of Notre Dame's large logic group, because I already enjoyed talking about math with them and admired their work. I did not have high hopes for getting the fellowship, but I figured it would be good practice for learning how to apply for grants. Since Michael and I were both applying for academic positions, we blanketed the country with applications. I was thrilled to get the NSF Fellowship. Michael's best offers, however, were on the east coast far from Notre Dame in the Midwest. We decided (and fairly quickly regretted) taking on postdocs more than a plane ride apart.

Other than that crucial snag, my three years at Notre Dame were incredible. I soaked up the vibrant logic life there that spanned not just computability theory but also model theory. In particular, I worked on several projects that measured the computational power encoded in various algebraic and combinatorial problems, expanding my research horizons. I collaborated with not only my two fantastic supervisors but also some of the many visitors and graduate students. We made our "two-body problem" known, and we were extremely fortunate that Notre Dame was able to offer Michael a visiting position after my first year. Though no one at Notre Dame worked in his area, we were thrilled to be living together and soon after he began to seriously consider a transition to industry.

I also reconnected with my SMP network while at Notre Dame; I began visiting the program regularly, mentoring current participants, and receiving advice from more senior people there. Visiting SMP reminded me how much I enjoyed working closely with undergraduates in a smaller setting. With that in mind, I sought out positions at liberal arts colleges that valued and supported teaching and research. Happily, I landed my dream job at Wellesley College, outside of Boston. Wellesley College even let me delay coming a year so that I could finish the last year of my postdoc. The department also offered Michael a temporary position, which he took.

It has been quite the (decade plus) ride since I arrived at Wellesley. Working with many people on many different things during my postdoc made it easier for me to keep my research program growing as a new faculty member. I also love that the kind of "applied" computability I do allows me to dabble in other areas of mathematics. I have kept on "throwing my hat in the ring" by applying to various research grants funded by the NSF, Simons Foundation, and others. Amidst many rejections, I have been fortunate to get a handful that have supported my research and travel, including one that even let me conduct summer research with undergraduates!

I also continued to develop my mathematical network outside of logic, e.g., through the SMP community and Project NExT (a professional development program for new mathematics faculty members run by the Mathematical Association of America). Project NExT and later the pandemic led me to experiment with various active pedagogies (such as inquiry based learning, flipped classes, and standards based grading). Out of an interest to become a better writer myself, I began teaching a capstone writing seminar on explaining math to the public. Fellow SMPers created gatherings to mentor SMP alumni currently in graduate school, and I regularly get to work with SMPers on various professional efforts (including this book!).

Things worked out for Michael too. After one semester at Wellesley, Michael started working at a small software company. He realized he loved programming and became a software engineer after taking lots of free online computer science courses. Over the years, we had two kids Aaron (age nine) and Naomi (age five), who make life an even bigger adventure.

Though I did not have a step-by-step procedure for finding my professional path, here are some things that worked for me:

1. Try lots of things out, whether it is an REU, a course, or a summer internship, to see what you like (and do not!). When looking for an area of study, aim to find a dynamic subfield that fascinates you and also has a welcoming, collaborative community that you want to be a part of. But, in the same vein, do not narrow your options too early. Though everything worked out for me, exploring more research directions before committing to computability would have strengthened my mathematical foundation and clarified my research goals. I also wish I had explored industry opportunities more in college and early graduate school.
2. It is never too late to make a change. Even though I wish I had considered nonacademic paths earlier, I am very glad that I took the time to explore career options before completing my PhD and that my position now gives me the freedom to take on new challenges. Michael is also thrilled he changed paths after his postdoc.
3. Continuously develop your professional networks (along various axes and career stages) and actively seek out advice and help. I feel very lucky to have had two doctoral advisors and two postdoctoral supervisors and to continue to be part of so many professional communities, from SMP to Project NExT to the Association of Symbolic Logic and more.
4. Lastly, take (calculated) risks. Apply for that position, scholarship, or grant you do not think you will get. Or take on that intimidating but intriguing project or position. You may or may not succeed, but you will learn so much. And you certainly will not succeed unless you try!

9.2 Emilie Purvine: My Meander to and in a Nonacademic Career

Throughout middle and high school I excelled in math and enjoyed it. I had many mentors including family and teachers. When I started college at the University of Wisconsin–Madison I knew I wanted to study math, but since the only career in math I was familiar with was teaching, I planned to major in math education. I wanted to become a high school math teacher and inspire students like I had been inspired. As I progressed in my studies, the math classes kept pulling me in, making me want to learn more. Towards the end of my first year, before I even made it to the math education classes, I decided to switch to a math major. I still did not really know what careers there were outside of teaching, but figured there must be some, and I could always do a math education master's later if I still wanted to teach.

During my freshman year I was told about the Summer Mathematics Program for Women (SMP) at Carleton College. I do not remember exactly how I heard of it, but I am so glad that I did. Being from Minnesota I knew of Carleton and it sounded like a fun way to learn more math while being somewhat close to home over the summer. I applied and was accepted! SMP opened my eyes to so many different things. It provided me with another set of influential mentors and I learned so much that summer, not just about math but about networking and future opportunities. I am where I am today in large part because of the things I learned that summer. I took the step to go to graduate school; followed a nonacademic career path much like some we heard about from guest speakers that summer; continued on in combinatorics, one of the two classes we took; and still keep in touch with the huge network of women that I met who all loved math just like me! The influence of SMP went so far beyond that one summer back in 2003.

Following SMP 2003 I went back to college with renewed excitement for math. I spent the summer of 2004 at the Summer Program for Women in Mathematics at George Washington University. This was a program similar to SMP that provided a very similar set of outcomes: encouraging me to go to graduate school, pursue a variety of careers, and grow my network of female mathematicians even larger. It seems that I kept finding excuses to do math!

During my final two years in college I added some (paid!) math activities. First, I began working as a tutor and student assistant in calculus classes. As a student assistant I helped the TA with grading and walking around during recitation sections to answer students' questions. I had really enjoyed calculus in high school so it was nice to share my enthusiasm with my peers in college. It turns out I did like the education side of math after all, but maybe at a higher level than high school. The second activity I added was undergraduate research. There was a professor at Wisconsin who was starting a semester REU in the area of algebraic combinatorics. I was able to cite my experiences learning about combinatorics at SMP and SPWM and get a spot in his research group. It was a fantastic experience. We met twice a week and talked about problems, learned how to use Maple to help in our research, and then worked outside of those weekly meetings to solve interesting problems. The third, and final, thing I added was a summer internship at a small government contracting company outside of Washington, DC. It was so valuable to learn about a nonacademic career in mathematics through hands-on experience! It was challenging but rewarding to think about solving real problems using math and computers. In all of these new activities I could not believe I was getting paid to help people with math and do math research! While SMP and SPWM showed me *how* to get to graduate school, these additional activities in my final two years of college taught me *why*.

Following that summer internship, my career goal became applied research in the government sector. I knew that a graduate degree would help me achieve that goal. In the final year of undergrad I took the GREs and applied to around nine graduate schools that had diverse programs. I thought breadth in mathematics would be helpful in a government career. Given my interest and undergraduate research in combinatorics I also made sure they had good combinatorics professors. I was accepted into two programs and decided to go to Rutgers University, in large part because of the graduate student community I met during my visit. I felt completely welcomed by the current graduate students. I almost felt like I was already part of their crew! I attended a seminar and met with professors, one of whom became my thesis advisor down the road; the undergraduate research topic I had been working on happened to be aligned with his work. Later, a group of students took me out to dinner and showed me around town. This kind of welcoming community was what I needed to survive and thrive in grad school. I knew that I needed a group of people that would be willing to work together and play together and I am so thankful that I found it.

My five years in grad school flew by. I took written qualifying exams (passed on my second try), an oral qualifying exam, many reading courses, and worked on research in rational difference equations. I was awarded a three year graduate fellowship from the Department of Homeland Security in my second year which allowed me to spend more time on research (no teaching required while on the fellowship) and intern at Pacific Northwest National Laboratory (PNNL). I spent three summers working on applied math projects using partial orders to study databases at PNNL, while I spent the school year on my pure math research. I stuck to pure math for my dissertation with the hope that learning a breadth of knowledge and a research mindset would allow me to be more broadly impactful in a future career (spoiler alert: it has!).

When it came time to apply for jobs I focused on the National Security Agency (NSA) and national laboratories. (Aside: if these are not familiar to you I suggest you look into them. There are 17 national laboratories around the country that perform interdisciplinary science research for the government.) I was offered a postdoc at PNNL and a mathematician position at NSA. I decided to stick with what I was already familiar with and went off to Richland, WA, to join PNNL in the same group where I had my internship. I have been here at PNNL ever since. After a year and a half I transitioned from postdoc to permanent staff. After a couple of subsequent moves up the ladder I am now a Chief Data Scientist.

9.2.1 Lessons from My Career

During my 12 years at PNNL I have learned a ton. I will not go through everything, but I will pick a couple of things that I think are important to highlight. First, as a mathematician working on applied science, part of my job is learning as much as I can about the application. When I first started my job I had an expectation that I would be handed math problems, but that is rarely the case. It is far more common to talk with domain scientists, learn about their challenges, and identify a math problem lurking beneath the surface. The only way I have found to do that effectively is to learn about the science. I am by no means an expert in the power grid, chemistry, biology, virology, or cybersecurity (all applications that I have worked in) but I know enough to understand the challenges the team is trying to solve. I have also grown a network in which I feel safe to ask any questions I have about the application.

The second lesson I will highlight is the importance of communication. When I sit on interview panels now I of course screen for technical skills, but I often look equally for communication skills. Doing the math is one thing, but communicating its impact can be much more difficult. We have to be able to talk to managers, domain experts, software engineers, and other mathematicians. I have learned how to write proposals, papers, and presentations at varying levels of complexity, and I have learned that I really enjoy the communication aspect of my job. I seek out opportunities to teach non-mathematicians about how math can be applied to their work, or to teach undergraduates about interesting applications of math. This is how I found myself in project management, a role I never thought I would enjoy. Turns out, I really do!

I have also learned a lot of new math. Coming in with a background in combinatorics and discrete math was helpful. There was a lot of work in applications of graphs and network science. But at a national laboratory I have had to be flexible and follow the priorities of the lab and the needs of the government. In 2012, I started a journey into topological data analysis (TDA) after attending a Mathematics Research Communities workshop. I now work in applications of topology to discrete and combinatorial structures that arise when modeling real data. I have also learned about machine learning (ML) and have recently started to explore foundations of ML models using advanced mathematical techniques.

While my journey has had some twists and turns, and at times felt like a meander, it was really quite linear. In hindsight I have no actual regrets. If I changed anything then I would not be where I am, and I am quite happy with where I am. But a linear path is not the only way. It is ok to take more time if you are unsure about something. Explore options. Take classes that might not quite align with your work. The more breadth you have early, the easier it is to pivot later. You may not remember everything about the undergrad logic class you took, but somewhere in the back of your brain you learned it and can access it more easily to re-learn it later. Having (somehow) not taken topology beyond point set in real analysis, I have had to learn it all as I go. I have been lucky to have great (and patient!) collaborators, but it would have been easier if I had taken that core class.

Finally, I have learned that saying "no" to something does not mean saying "no" forever. I pivoted away from teaching early in college but came to mentoring students as interns and giving talks on my research and related concepts at conferences and university seminars. It is not lesson planning and grading, but it is still teaching. Changing your focus is ok! It gets harder as you get older, but it is still doable. Do not stick with something because "past you" thought it would be awesome. If it is not, it is not. Believe "present you." Give it enough time to be sure, but do not be afraid to make a move. I have never made a hard pivot, but I definitely did not see national labs, project management, or TDA coming and I am incredibly glad I took those opportunities when they came my way. All I know about what is next for me is that I do not really know what is next. What I hope is that I can continue the broad themes of mentorship, mathematics, and learning.

9.3 Erin McNicholas: Finding Your Voice, Your Identity, Your Place in Mathematics

9.3.1 Carried Along by Other People's Currents

As an undergraduate and graduate student, I was highly motivated by a desire to please others. In college it was my professors and parents, and in graduate school it was my professors and partner. I tried very hard to be everything to everyone. Finding intrinsic motivations, understanding how my efforts were ultimately for me, and taking a chance by just being me were all skills I had to develop in graduate school and beyond.

Since I did not have a clear sense of myself or what my goals were, I was propelled forward by others' expectations. In fact, I would have been directionless without the pressure to live up to what I thought others expected of me. I found beauty and interest in math and physics, but I majored in those disciplines primarily because they gave me a way to please people whom I deeply admired. I started a graduate program in particle physics because I was encouraged to do so. I chose a graduate program a few years later based on where my partner wanted to relocate. Reflecting on these decisions now, I realize they were not well-thought-out plans. I was floating on the currents of others' dreams or wishes for me.

In my experience, a people-pleasing approach to one's vocation runs its course. Making choices to please others may bring them pride and happiness in the short term, but it wears off and you are left with the commitment, the workload, and a need for some other—more *intrinsic*—motivation. I felt this need in my first year of graduate school in physics. While I love physics, I am not an experimental scientist. It took me a while to realize it, but what I actually loved about physics was the symmetry laws, the underlying mathematical structures. Graduate school in physics was also my first experience of feeling part of a noticeably small minority. Having had such a supportive undergraduate experience, I was not prepared for the sense of isolation and discomfort of being the only woman in a lab. Back then, I did not know how to advocate for myself or how to push back against what was a pretty hostile environment. I ended up leaving after just a semester—disappointing many of the people I had hoped to please by going to this program in the first place.

Through this experience I learned that *it is ok to leave an environment that is not supporting you!* It was not easy. At the time I felt completely lost. I had to move back home and felt under tremendous pressure to take whatever job I could find in the least amount of time. Suddenly, the future that had seemed so clearly laid out before me vanished. I felt I had failed: failed to see it through, failed to be strong enough, failed to be smart enough, failed to be impervious enough.

Fortunately, I eventually found my footing. I interned with the American Physical Society, attending congressional hearings, helping lobbyists, and writing short pieces for their newsletter. Technically, I should not have been eligible for this internship. I took a chance by applying and writing a humorous cover letter that suggested it was either this internship or I would have to become a roadie for a country singer and that I did not look good in leather fringe. Ultimately it was the cover letter that got me the position and more writing responsibilities than I might have had otherwise. It was also a good lesson that while it often feels risky, being your true self can set you apart. While I loved the writing aspects of the position, I learned I did not enjoy lobbying and missed Oregon.

I got a job as a metrologist (someone who ensures the standards of weights and measurement used in science and commerce are maintained and accurate), moved into my own place, and met my future husband. This job afforded me a lot of time to contemplate what my goals were. I found I missed my math classes. I started sneaking my undergraduate textbooks into the office so I could reread them during downtimes. As an undergraduate, I had been a participant in the first year of SMP. I kept thinking of the impressive and brilliant women I had met there, the teaching assistants who were thriving in their graduate programs, the impressive faculty who were making great contributions in mathematics and also taking time to challenge, push, and encourage women like me. My experience at SMP planted a seed. I wanted to be part of *that* community. Fortunately, Deanna and Steve, the directors of SMP, had worked to foster a sense of community that extended beyond the four weeks of the program. As I surreptitiously read *Contemporary Abstract Algebra* in my cubicle, I was getting updates about SMP alumni who were completing their master's degrees, passing their qualifying exams. Within a few months, I knew I wanted to go back to graduate school, this time in mathematics.

I am sure you are thinking that the second time around I did more research and was much more thoughtful in my decision of which school to go to. Unfortunately, I did not. Life lessons can take a while to sink in. I chose a program largely based on the needs and interests of my partner. But I was a different person this time. Knowing I could leave, that *it is always possible to change course*, gave me the strength to navigate situations I had not known how to navigate two years earlier. I also had the support of the SMP community. To supplement my graduate school stipend and explore interests outside my thesis area, I interned with the cryptology group at Sandia National Laboratory starting in the summer of my second year. One of the undergraduate cryptology interns I met during my time at Sandia had also taken part in SMP, and I was bolstered by seeing the positive impact the program had on her as well.

I was fairly certain I wanted to be like the faculty mentors I had met at SMP. I wanted to do research, but I also wanted to have close relationships with students. I took advantage of opportunities in graduate school to improve my teaching skills. I took classes from the school of education; I taught as many different undergraduate courses as the department would offer me; I went to optional workshops and training sessions. When I graduated with a PhD in Applied Mathematics 6 years later, I was far more intentional in looking for jobs than I had been in either grad school search. I considered where I wanted to live and what kind of position I wanted to take. Most of all, I considered what kind of department I wanted to be part of. I was fortunate to find my place at Willamette University in a department filled with innovative, supportive, and inspiring professors. Over the last 16 years, I have been able to collaborate with my department colleagues on both research and teaching projects. And I have found that *when you are supported and valued, you can flourish in a way that is just not possible under other circumstances.*

9.3.2 Finding Your Niche, Finding Your Talents, Getting Out of Your Own Way

It was not until years after graduate school that I learned to value the variety of approaches and forms of intellectual contributions within mathematics, my own contributions and approaches included. I think we have a tendency as a mathematical community to value independent genius over other archetypes. We prize that individual who seems to live and breathe mathematics, who arrives upon the right solution or insight instantaneously and independently. I certainly met many students in graduate school who matched this stereotype better than I did. I felt that to be successful, I had to somehow become more like them. I thought I had to hide the fact that I had nonmathematical interests, that I was not always satisfied to work on my research alone, and that at times I felt completely lost. I felt I had to constantly project an image that was more confident, more competent, and more passionate about the subject than I often felt.

It took me a long time to learn to get out of my own way—to be honest with myself and others, to ask clarifying questions, to admit what I do not know, and to ask for help when I need it. This humility is still a struggle. There are certain collaborations where I feel completely able to be myself and others where I still feel compelled to hide my ignorance or uncertainty. I see this same struggle reflected in some of my students—they think they should understand everything on their own, too afraid to admit when they do not, lacking confidence in their own abilities and contributions.

What I have learned is that *I am a much stronger and more productive mathematician when I am open, honest, and true to myself.* I clear up my confusions more quickly; I offer perspectives and ideas even when I am not confident in them, giving my collaborators the chance to challenge my ideas and help refine them. The end result is a much more efficient and enjoyable form of research and collaboration.

I have come to realize that I am not the best at making continued and steady contributions. Like most people in academia, I am constantly overcommitted. My teaching responsibilities, personal responsibilities, and general varied interests put demands on my time and pull at my attention. What I am good at is bursts of productivity and focus. This realization has shaped the collaborations I seek out. I have successfully found projects that allow for progress that ebbs and flows. I have sought out opportunities offered by various mathematical institutes (such as AIM SQuaREs and Collaborate@ICERM) to meet with collaborators in person and work intensively for weeks at a time. I have learned that I am good at seeing patterns, finding algebraic framings, and spotting holes in arguments, but not as good at posing questions or boldly taking first steps. I learn quickly but work methodically, and I only really enjoy math when I have the opportunity to share it with others—my students and my coauthors.

I have had the great fortune of working with a diverse collection of talented mathematicians. I have learned so much from each of these individuals, from undergraduate co-authors to more senior collaborators. I have learned to appreciate the diverse approaches we bring to a project—the way our combined perspectives result in brave leaps of logic and methodical attention to detail, the new questions alongside connections to established results, the pragmatic planning and whimsical wondering. It might be a well-timed observation by one collaborator that leads to the "aha!" moment for another. It might be the question posed by one individual that gets answered by several others. This is one of the most beautiful things about authorship in mathematics—that contributions cannot easily be disentangled and quantified and that the finished product is more than the pieces. This is reflected in the discipline norm of listing co-authors in alphabetical order. Co-authorship in mathematics is choral work more than a collection of solo artists. *Finding harmony takes trust, honesty (with yourself and others), and an openness to the ideas and talents of others that comes from recognizing that you have a place here in mathematics.*

9.4 Abra Brisbin: Charting a Path to Grad School and Beyond

9.4.1 How Did You Decide to Go to Graduate School?

I decided to apply to graduate school before I decided on applied math and statistics as my field. This was, in part, thanks to my undergraduate professors at Carleton College, who encouraged me to consider graduate school. I admired their teaching and aspired to follow in their footsteps. The summer after my junior year, I participated in the George Washington University Summer Program for

Women in Math (SPWM), a program similar to SMP, but aimed at juniors rather than sophomores. There, I took intensive courses focusing on just math, a different experience than the varied course load I took during the academic year. Realizing that I enjoyed this experience helped me affirm that graduate school was a good choice for me. SPWM also helped me envision myself as a graduate student, because the colloquia and panel discussions gave me so many opportunities to ask questions of successful women who had taken diverse paths to their PhDs and beyond. Simply living and working with a large group of other students who were equally passionate about math helped solidify my enthusiasm and sense of direction in my decision to attend graduate school. I think that this is very similar to what most student participants experience in SMP.

9.4.2 How Did You Choose Your Subfield?

When I was initially thinking about graduate school in my junior and senior years of college, I felt torn about whether to go into pure or applied math. I loved applications of math to biology, chemistry, and physics, having taken many science electives in college. But I also loved the courses I had taken in number theory and abstract algebra. An independent study in probability models of DNA helped make up my mind. I was fascinated by the intersection of probability and genetics, but I knew that many of my friends were not; most of my friends who were applying for graduate school had chosen programs in pure math. Given that I had similar levels of interest in both pure and applied math, I thought I might be able to do more to help advance the sum of knowledge in the field that seemed to be less popular among my peers.

I am happy I chose statistical genetics. Taking classes in math, statistics, computer science, and biology gave a sense of variety to my graduate coursework and helped me get to know students from other disciplines. Learning about what problems were important to biologists gave me inspiration for what questions to investigate in my own research.

I found that I really enjoyed the computer programming aspects of statistical genetics research. Designing the structure of a program—and debugging it—had a puzzle-solving feel similar to what I love about math. I felt that "when my code isn't working, I can't go home until I fix it; but when it is working, I don't want to go home!" This attitude led to plenty of late nights in the lab! (Of course, it is also an exaggeration; sometimes the best way to solve a stubborn bug is to get some sleep and approach the problem from another angle. One time I even found a major bug by *printing out* my code on about 20 pages of paper, taking it home, and reading through it with a highlighter and a colored pen.) I also liked that doing simulations gave me a clear entry point into making progress on problems that I would not have known how to get started on, if I had been limited to the pure theory on paper. In college, I did not really anticipate how much I would enjoy computer programming; I had used Mathematica a bit in some math classes, but I had never made time to take computer science courses in college. However, it makes sense in retrospect, as I did spend plenty of my free time playing with LOGO and BASIC back in elementary school.

Since finishing grad school, I have also enjoyed my choice of statistical genetics. Statistics is a great field for being able to collaborate with scientists in other domains and getting to learn about interesting topics that might be unrelated to your original area of application (in my case, genetics). I have gotten to learn about linguistics, hip surgery, adoption trends, and more through collaborations making use of my experience with statistics. Statistics and computer programming also gave me an entrée into data science, which has grown in popularity since I finished grad school. I have learned a lot about machine learning and got to work with colleagues to develop two brand-new courses in the online University of Wisconsin Master's in Data Science program, which was a lot of fun.

If you are genuinely undecided about pure versus applied math, like I was, I should mention two tangible advantages to applied math. First, grant money tends to be more available in biology than in pure math, because biology research can be supported by the NSF or the NIH. In graduate school, my advisors had grants for their computational biology research. This meant that they were able to give me research assistantships during about half of my semesters in graduate school; the other semesters, I did teaching assistantships. Being paid to work on my dissertation some semesters definitely made it easier to earn my PhD in 5 years. Some of my friends in pure math took 8 years to graduate, because grading or teaching every semester took up so much of their time. For some of them, this was just fine—they were at a stage in their lives when they were happy to focus on learning and doing research, and they did not care that they were still only earning grad student money. But others started to get tired of grad school by the end, even though they had come in excited to spend however long it might take. (My co-editor Erin would like me to remind you that the plural of anecdote is not data. The median time to a PhD in math and statistics is 5.3 years,[1] and only 33% of new math and statistics PhDs come from statistics, biostatistics, or applied math departments.[2] This means that plenty of pure math grad students finish in 5 years or less! But I think it is important to keep in mind that your goals can change. Even if a long, leisurely exploration of all the nooks and crannies of your research area seems great at the start of graduate school, it might not be what you want by the end.)

The other tangible advantage to applied math and statistics is slightly better job prospects compared to pure math. According to a 2018 AMS survey, the unemployment rate of new PhDs in statistics was 2.7%, compared to 5.9% for new PhDs in math. This is probably partly because it is easier for statisticians to explain how their education has prepared them for jobs in industry. Higher mean salaries for new assistant professors in departments of applied math and statistics[3] compared to math departments also speaks to the demand for these fields. These are

[1] https://ncses.nsf.gov/pubs/nsf24300/data-tables

[2] https://www.ams.org/profession/data/annual-survey/2018Survey-NewDoctorates-Report.pdf

[3] https://www.ams.org/profession/data/annual-survey/2019Survey-FacultySalaries-Report.pdf

certainly not reasons to choose applied math or statistics if what you truly love is pure math. But if you love both pure and applied math, this may be a valid reason to tip the scales toward applied math. (A word to the wise: Do not get too hung up on the AMS salary tables. It is definitely worthwhile to negotiate when you receive a job offer, and these tables provide data that can be helpful in your negotiations. But remember that these values represent a wide range of institutions with different financial resources, in cities with different costs of living. If you get too focused on thinking you "deserve" a certain salary, because it is the mean or median for a given type of institution, there is a good chance you will end up dissatisfied—whether that is because the job you really want will not meet your salary demands, or because you tricked yourself into taking the highest-paid option, regardless of how expensive the city is to live in, or what other advantages or disadvantages the job has. There is more to a job than just the salary. If a job you are interested in will not increase its salary offer, look for other dimensions you can negotiate that would help you be happier and more successful in the job, such as support for conference travel, a reduced number of new course preps during your first year, or the opportunity to teach in the summer to augment your salary.)

9.4.3 SMP

I was a TA at SMP in 2004, the summer after I graduated from college. For me, this was both a learning experience and an opportunity to "give back" to the mathematical community by encouraging younger students. The other TA, Andrea Frazier, was already in graduate school, and she was a role model for me. Just as important as my experience that summer was the sustained sense of community that SMP has provided. Seeing SMP alumnae—from SMP 1995 to 2014 —at the Joint Mathematics Meetings every year has allowed me both to receive and to provide mentorship with an extensive network of mathematical colleagues, which in my experience has been more effective and reliable than the various one-on-one mentorships to which I have been assigned at different points in my career.

9.4.4 Advice to Current Undergraduates

Unfortunately, SMP and SPWM have both lost their grant funding. For other students who are considering graduate school in math, I would recommend immersing yourself in as many math-related opportunities as you can find in college. I did an REU (Research Experience for Undergraduates) in biomedical engineering and spent a semester studying abroad through the Budapest Semesters in Mathematics program. I worked as a tutor in my college's math lab and competed on the math team. All of those experiences helped me learn more about math and about what I wanted out of my career. However, your life does not have to be 24/7 math: Between math classes and extracurriculars, I also took electives in lots of other subjects, played in my college jazz band, and performed improv theater. Wanting to have a life outside of math does not mean that grad school is not for you!

For students who are specifically interested in statistical genetics, I recommend taking probability in college. Real analysis is also a good choice, to prepare for measure theory. I think that ideally, your course schedule would include a good mix of statistics courses. I was kind of prejudiced against statistics in college; in high school, I had taken a course in AP statistics that was only "ho-hum," so I avoided it in college. In grad school, I found out that not only is statistics useful for understanding genetics, but it also involves a great deal of mathematical theory and puzzle-solving, which I enjoy; I need not have avoided it. I also recommend taking biology courses, to get a better understanding of the concepts you will be modeling with mathematics. I took 4 biology courses in college, including one on genetics.

I never found time to take any computer science courses in college; I was more interested in other courses. In grad school, I took courses on programming in C and on "Unix tools" (basic shell scripting and command-line functions I could use in Linux) and picked up R through using it in several of my courses. In my research, I do a lot of computer programming, mostly in R. So, I think that taking computer science courses in college would be beneficial, but I am proof that it is not absolutely necessary. While the ideal situation would be to take courses in the programming language(s) that are used the most in the field you are interested in, the course I took in C taught me a lot about logical programming structures and strategies for thinking through the structure of code that made it easier to learn other languages later.

Lots of people advised me that good writing skills would be important in graduate school, and I agree with that advice. What no one mentioned was that graphic design skills also come in handy for designing research posters for conferences! If I could go back in time, I would add a class on graphic design in PowerPoint and Illustrator to my college experience, rather than having to pick things up on the fly when I was trying to emulate other people's formats for my first poster in graduate school.

9.4.5 Applying for Graduate School

I took a year off from taking classes between college and grad school, and I was glad I did. It meant that I was not trying to write applications and visit schools at the same time as I was working on my senior thesis, and I did not have to figure out where to take the GRE in Budapest while on my study abroad program. It also helped me ensure that I would not feel "burned out" on taking classes as I was starting graduate school. I was able to find a job as a "fifth-year intern" teaching assistant at my college, which was good preparation for being a teaching assistant in grad school.

I took the general GRE and the math subject GRE in the fall of my gap year. I took some practice exams for the general GRE, but I spent much more time studying for the math subject GRE. I bought a used math GRE test-prep book from one of my friends who had taken the test the year before and worked on problems from it throughout that summer and early fall. I focused on reviewing calculus and linear algebra, since those topics make up the bulk of the test and had also been the longest ago among my college classes.

Early in the fall of my gap year, I started researching grad schools. The task seemed a bit daunting at first. I paid for access to the US News & World Report list of the top 100 graduate schools in math, just so I would have an initial list that I could pare down. Then I explored the schools' websites to identify which of their graduate programs best fit my interest in statistical genetics: at some schools it was statistics; at others it was applied math; and at still others, it was computational biology or genome sciences.

I spent hours tailoring my statement of interest to each of the graduate schools I applied to. I tried to avoid rehashing my resume by drawing connections between the mathematical things I had done in college and how they had prepared me to contribute to, or benefit from, specific courses, seminars, or activities I had read about on the graduate program's website. (For example, "Based on my experience as a teaching assistant for the Carleton College Summer Math Program, I look forward to the possibility of contributing to your department as a Graduate Student Instructor. Your course on the Teaching of Mathematics would be a valuable next step in improving my communication skills, and an important opportunity as I work toward my goal of becoming a professor.") I took a similar approach to cover letters when applying for jobs later. Now, having served on four recruitment committees for tenure-track positions at UW-Eau Claire, I can attest that most people do not bother to tailor their cover letters, except to mail merge the name of the school they are applying to (if that). Taking a more detail-oriented approach really helps your application stand out. If you are not taking a gap year and have less time to do this, even a couple of sentences tailored to each school can be beneficial. Out of hundreds of possible choices, you chose to apply *there*; tell the admissions committee why. (In my opinion, it is OK if one of the reasons is "I have family in the area" or "I'd like to live in your city," as long as you also give a more academic reason, too. It should be more specific than just "your school's reputation for excellent academics;" again, there are hundreds of schools with excellent academics—what makes this one stand out to you?)

9.4.6 How Did You Choose Which Graduate School to Attend?

I considered schools all over the United States, but with a preference for areas where I thought I might like to live. I focused on small graduate programs (those with a total of around 40 students), because I enjoyed the close interactions with my professors that I had had in college, and I wanted a similar environment in graduate school. I consulted with my professors about their thoughts about some of the schools, and I sent emails to current students in those programs to ask what they thought about the program. Initially, I emailed alumni from my college and former participants in SMP and SPWM—people with whom I had some connection, even if I had not met them. Then I filled in the gaps by emailing random students, and occasionally professors, from the graduate programs' websites. I found that even the strangers were usually very willing to respond, as long as I addressed them directly so it was clear that I was not spamming the whole department. Responding to prospective students is a productive form of procrastination! Based on a combination of the program websites, my professors' recommendations, and emails with current students, I winnowed my list of graduate schools down to 8 and sent in my applications.

I was accepted to 7 programs, and each of them offered to help pay for me to visit the school, either individually or as part of an accepted students' weekend. (The most biology-oriented program actually had a pre-acceptance weekend, which was primarily about encouraging us to attend, but I guess it left the school the option of deciding not to accept someone if they started shouting and breaking things in the middle of the welcome reception.) I was glad I visited all the schools, because it really changed my opinion of some of the programs. At one school, there was a sense of stress and competition among the students about the qualifying exams, which I had not been able to detect from their website. I knew that was not for me.

I decided to enroll in the Center for Applied Math (CAM) at Cornell University. I liked that there were several professors there who worked on computational biology and applications of probability to genetics, so I would have options for potential advisors. (This turned out to be very important.) I liked the fact that the Center for Applied Math did not have a set of core courses that all students were expected to take; just a set of distribution requirements that gave me a lot of flexibility to choose classes that I thought would be relevant to my field of interest. I could take classes in multiple departments—math, statistics, operations research, and biology. (For fun, I even took a class on Introduction to Wines.) I liked that the current students assured me that failing one's A exam (the admission to candidacy exam, where students presented their initial research progress and their plans for their dissertation) was extremely rare; your advisor would not let you schedule it until they believed you were ready. This, along with the lack of qualifying exams, seemed to contribute to a sense of cooperation, not competition, among the students. When I went out to dinner with a group of current students, they all had some hobby that they did outside of doing math, and they could talk to each other about things other than math. One person was taking a pottery class; another was on an intramural soccer team. I could attend Cornell and have friends and life beyond my thesis.

In some ways, I got lucky in my choice of graduate schools. The stipend for research assistantships and teaching assistantships comfortably covered the cost of living in Ithaca (at least at that time), so I was able to rent an apartment without roommates, eat at restaurants with friends about once a week, and even save some money for after graduation. This was lucky, because coming almost directly from being an undergraduate, in a different part of the country, I did not really have context for what stipend was "enough"—any stipend, on top of getting *free* tuition, seemed like a great deal. I am also glad I chose a program in applied math instead of genome sciences, because my goal was to become a professor in a math department, not a biology department. Depending on the institution and how the qualifications in the job ad are worded, the official field of your degree can make a difference in whether your job application makes it past the first cut.

Finally, I benefited from attending graduate school in a different part of the country than where I had grown up and gone to college. I enjoyed living in Ithaca, and knowing that was valuable information for deciding where to apply for jobs. My anecdotal experience was that I got better response rates from applications for tenure-track jobs on the East Coast and Midwest than in other parts of the country; I hypothesize that seeing that I had attended grad school and college in those areas made it more credible that I knew what I was talking about when I said I wanted to work there.

9.4.7 Advice for Current Graduate Students

My graduate program assigned me an advisor to help me choose classes for my first year in graduate school, but her research area was very different from my interests. I got good advice about classes from the experienced grad students, but I also wanted to get the perspectives of some of my potential dissertation advisors. I emailed two of them and politely asked to set up meetings to discuss my course schedule for the upcoming semester. I met with both of them, plus my assigned advisor, and put together a schedule. This turned out to be a fantastic strategy to start to get to know my prospective advisors—far less intimidating than asking to do a reading course with them. I got their advice before the start of spring semester, too. That was enough to make it clear to me that I had a big personality mismatch with one of the two potential advisors. (To be honest, I had an inkling that we would not be a good match even based on the emails we had exchanged when I was choosing which graduate school to attend.) Fortunately, I was at a school where there were other advisors I was interested in working with. I asked the other professor I had been meeting with if he had any projects I could work on over the summer. He asked me some questions about my interests and then invited me to work on a simulation of natural selection in the computer lab shared by his students and the students of another professor, who was newer to the university. That project did not end up becoming part of my dissertation, but both professors in that lab liked my work, and I liked working with them, so at the end of the summer, they asked if I would like to officially join the lab. Both of them were my co-advisors for my dissertation.

For me, one significant difference between classes in grad school and undergrad was that in a PhD program (especially one like CAM that has distribution requirements instead of a set of core courses paired with qualifying exams), your course requirements are really about what your dissertation committee—your advisor(s) and a small number of other professors—think you need to succeed in your research and be prepared for after you graduate. When I asked one of my advisors which course I should take to fulfill my "algebra" distribution requirement, he replied, "I don't think any of these will be that helpful to you... let's have you take this course on Computationally Intensive Statistical Inference instead." It was a great class, and all three of the advisors on my committee signed off that I had satisfied my distribution requirements.

One day in my first year, on my way back to the CAM grad student office after class, I ran into Deena Schmidt, another CAM student (who has a chapter in this book) with similar research interests to mine. She was on her way to a "journal club." I did not know what that was, but I tagged along. It turned out to be an informal class in which students and professors met to discuss journal articles on a selected topic. Typically in a journal club, a different person is responsible for giving a summary of each week's paper to get the discussion going. I was incredibly nervous when I gave my first journal club presentation, and sure enough, the professors in the room had spotted things about the paper that had gone way over my head. But hearing how the more experienced people in the room reasoned through each paper helped me better understand what aspects of a paper I should focus on, and reassured me that I was not the only one who sometimes struggled to understand a paper's meaning. I was amazed to learn that we could even critique the methods of published papers. Gradually I gained confidence and the background knowledge I needed to understand the papers in my field and even critique their methods sometimes.

In addition to taking classes and attending journal clubs, I watched bulletin boards and my email to find out about colloquia that interested me in math, engineering, and biology. I liked getting a taste of research topics in a variety of fields (not unlike this book!), and the speakers would sometimes mention a technique or paper that would give me an idea for my own research. Less formally, I also observed some public speaking techniques that I did and did not want to emulate. Hearing about the speakers' careers—at different types of academic institutions, in industry, and in government—gave me a taste of my own possibilities. And getting to know grad students in other fields over snacks before or after the colloquia was helpful for forming study groups; feeling a sense of connection (we are all in this together!); and gaining perspective on what is normal for your school, what other departments do that you would like to emulate, and when you should be grateful that "at least my department does not do *that*!" Volunteering for Expanding Your Horizons, a STEM activity day for high school and middle school girls and non-binary students, was another way I got to know grad students from other fields, as well as pay forward the mentorship I had received through SMP and SPWM.

Grad school is definitely challenging at times. For many of my friends in the pure math program, the first year was the hardest, because that was when they were taking their required "core courses"—which substituted for qualifying exams and were so intense that they were advised to take only 2 courses per semester. For me, the second year was more difficult, because of the new challenge of balancing research in my new advisors' lab with my coursework and TAing. Several things helped me get through the rough patches. First, forming study groups with classmates made it easier and more fun to do our homework. It was really helpful to be able to talk through ideas and have a peer challenging me to justify my reasoning, or give me a hint about a problem I had not solved yet. And it felt great to have my friends compliment and thank me when I was the one who made a breakthrough on a problem. Initially, I formed study groups by asking people who sat nearby me in different classes, which included students in CAM, engineering, and biology. After we found a core group of people who worked well together, we would discuss our course schedules for the next semester so we could continue to work together.

Another resource that helped me do my best in challenging graduate courses was... undergraduate courses. Since I had not taken statistics in college, I took a senior-level undergraduate statistics course for credit while I was a graduate student to help prepare me for graduate-level statistics. I also audited an undergraduate computer science course and bought the book for an undergraduate Algorithms course to read alongside the book for my graduate Algorithms course.

Throughout all of this, having a strong support network helped me stay mentally healthy. Doing swing and contra dancing gave me exercise and a group of friends from outside of math and reminded me that I was a whole person, not solely a grad student. Even though I was over a thousand miles from my family, I stayed connected to them by calling them most nights on my walk home from the lab (the fact that they were one time zone *earlier* than I was definitely helped). One of my friends got a referral for a therapist through Student Health Services and swore that it was incredibly helpful to her mental health while writing her thesis. It was also helpful to have a clear reason for *why* I had chosen grad school—to become a professor and follow in the footsteps of my college professors, who were excellent teachers—to keep me motivated when I was frustrated or doubting myself.

During my time in graduate school, I TA'ed a range of courses, from Calculus I to Quantitative Genetics. One semester I got to design my own course for the Senior Math Seminar at the local high school. It was a lot of fun to work with a small group of really motivated students and develop my own curriculum. (I chose Introduction to Probability; one of my friends did Group Theory and Rubik's Cubes.) I also volunteered for an outreach program in which graduate students designed 8-session minicourses that we taught at middle schools and high schools. (I did "Probability in Board Game Strategy " and "Computers, Math, and DNA.") The experience of teaching and TA'ing for a variety of courses, including ones where I could plan my own lessons, let me experiment with different approaches to active learning and gave me something concrete to write about when it came time to construct a teaching philosophy for job applications. I do wish that I had the chance to TA for the same course and professor twice; that might have increased my chances of being nominated for a TA award, especially if I had been more upfront about asking the professor to observe one of my recitation sessions and consider nominating me.

My thesis consisted of three papers combined into one document, one paper per chapter, with an introduction and conclusion. I submitted the first paper to a journal before I wrote my thesis; the other two I submitted after I graduated. Chapters 1 and 2 dealt with different approaches to solving the same problem (using genetic data from families to locate genes for ordinal traits, such as diseases that can be mild, moderate, or severe), while Chap. 3 addressed a different problem (identifying the ancestry of different portions of the DNA of people with mixed ancestry). The best advice I got about writing my thesis was to write for 15 minutes first thing every morning when I got into the office. Fifteen minutes was short enough that it seemed doable, even on days when facing my thesis seemed monumentally stressful. And often, once I got started I would get into a groove and be able to keep writing for a few hours. Even if I needed to quit after the 15 minutes were up, those 15 minutes were enough to keep my thesis fresh in my mind, so that when I came back to it the next day I did not need to waste time reminding myself where I had left off. I have used this strategy a few times since then when working on papers, and it has helped me out then, too. Note that in this context, "writing" refers to any activity that directly advances your progress on the thesis: actually typing new sentences, rephrasing things you have already written to make them more clear, typing a bullet-pointed list of things you want to remember to write about, looking up citations for specific things in your thesis, or doing analyses or making graphs that you are going to include in your thesis as soon as they are done. What it does not include is course prep or grading, or analyses that seem super interesting, but that you are not sure how—or if—they fit into your current chapter.

Remember that potential advisor I considered during my first year, with whom I had the personality conflict? I asked him to be on my thesis committee, as I needed a third professor in addition to my two advisors, and his research really was related to my research and interests. When it came time to schedule my A exam (admission to candidacy exam), I emailed all three of my committee members and got lists of which days they were available. One of my advisors was doing a lot of traveling that semester, so I was limited in my options. All three of them appeared to be available on a particular morning, and one of my labmates had his B exam (dissertation defense) that afternoon. Consolidate all the stress and celebration into one day, right? Wrong. My third committee member wrote a terse email replying that because my labmate's B exam was that afternoon, he "wanted to get some work done in the morning." Ouch. Not only was this a hassle, because he had not said anything about my labmate's exam when I had asked him for his list of time conflicts, but it also implied that mentoring us graduate students did not count as real work. I vented about this privately to the people in my support network, and then sent a conciliatory email suggesting an alternative date. ("It certainly is challenging to find a time that works with four people's schedules, isn't it?")

I got through my A exam OK, but as my B exam grew closer, my third committee member began sending more and more signals that he did not want to be on my committee. He suggested that I should find a committee member who could advise me about algorithms (presumably, instead of him). He talked about how his summer travel would interfere with my B exam (even though his travel was in July and I eventually took my exam in May). I do not remember what the final straw was. But in September, eight months before my B exam, I broke down crying in the graduate director's office. He helped me find a different committee member, and it all worked out okay. I found out through the grapevine that my old committee member had pulled similar tricks to persuade several other grad students to "let him off the hook" for serving on their committees. This experience really drove home the importance of finding mentors who make you feel welcome, and it made me very glad I had chosen a grad school where I had other choices of advisors in my research field. I was proud of the diplomacy I had used in my emails to keep my committee together for my A exam, and I was also proud that I had advocated for myself and found a different committee member for my B exam.

9.4.8 After Graduate School

I had started graduate school with the goal of emulating my college professors, who were excellent teachers at a school that valued both teaching and research. While I attended several talks and panel discussions about careers in industry and government, out of curiosity, my goal of becoming a professor persisted. I knew that doing a research-oriented postdoctoral fellowship (or postdoc, for short) first would help me gain independence in my research. That would be valuable if I ended up at a smaller school where my colleagues' research areas were very different from mine. As a side benefit, this also gave the economy two extra years to recover after the 2008 subprime mortgage crisis. (I finished my PhD in 2010.)

In the summer before my fifth year of graduate school, I attended the Joint Statistical Meetings and participated in the Employment Center there. I had interviews with several employers, but the one I clicked with was a statistical geneticist at Mayo Clinic. I wrote the interviewer a follow-up thank-you email; she invited me for a campus interview and then hired me for a two-year postdoc. The postdoc was a great experience for me, because I got to see a more medical perspective of my research field. Getting to focus more on research for a while was also refreshing; I think that for me, heading straight from writing my thesis to teaching a full load of classes in a tenure-track job would have been a recipe for burnout. Like taking a gap year between college and grad school, the postdoc also allowed me to focus on applying for tenure-track jobs when I was not writing my thesis.

During my postdoc, I worked on some new papers and continued working to publish the papers from my thesis. (Chapter 3 did not get published until after I started my tenure-track job. But that was OK, because it meant I could include it in my binder when I was applying for tenure.) I also asked to be able to get some teaching experience, to prepare for a future tenure-track job. My advisor arranged for me to teach the first half of Introduction to Biostatistics for the Mayo Graduate School, one semester during each year of my postdoc. I asked one of my advisor's colleagues to observe the course and write a letter of recommendation about my teaching for my applications for tenure-track jobs. Although my postdoc was at a primarily research-oriented institution, my advisor's colleague was an excellent, dedicated teacher, so it was really valuable to have her first-hand observations (from two class periods which she attended without telling me in advance) rather than a generic letter based on just my student evaluation numbers.

Doing a postdoc at a medical center felt lonely at first. Unlike in graduate school, I was not starting at the same time as a cohort of other trainees. Most of the coworkers I saw on a daily basis were at a different point in their lives and careers; after work, they wanted to go home to their families, not hang out. To make friends, I had to put energy into connecting with people through hobbies. I joined a walking club through MeetUp.com; a friend of a friend from my college swing dance club introduced me to rock climbing; a random conversation at a street festival led to joining a monthly book club in a town an hour away; I joined the local Unitarian Universalist church and became a dedicated member of its LGBTQA Committee. It took a few months, but I eventually re-established the social network I needed to feel like a whole person. The fact that my postdoc was for only two years helped me feel comfortable volunteering for more things, because I knew that I was not signing up for life; if I ended up feeling overcommitted, I would have a natural "out" at the end of two years.

During the second year of my postdoc, I applied for mostly tenure-track jobs. I eventually was offered, and accepted, a position in the Department of Mathematics at the University of Wisconsin-Eau Claire. This department is a great fit for me in terms of valuing teaching, supporting lots of research with undergraduate students, and having a collegial department. I was fortunate to receive a job offer less than two hours from my family; when my mom's health failed, I was enormously glad that I chose this position instead of the higher-paying one halfway across the country.

During my first year at UW-Eau Claire, I participated in Project NExT, a professional development program for early-career faculty run by the Mathematical Association of America. This was a great opportunity to get to know mathematicians at other schools and learn about a variety of teaching techniques I had not tried before.

Since joining UW-Eau Claire, I have diversified my research to include the Scholarship of Teaching and Learning and applications of data mining to a variety of topics chosen by students, such as adoption trends or local crime statistics. I really like that my experience with statistics and computer programming has given me flexibility to explore many different fields of application.

Most of my research now takes place through collaborations with undergraduate students, so I have come to value topics that are accessible and likely to pique a student's interest. I like to think that I have become good at noticing questions or issues in my environment where I have a chance to make things better using my analytical skills. For example, I became chair of my department last year. Instead of continuing to schedule math classes by hand in Excel, I am working with an undergraduate student to use linear programming in Python to optimize the course schedule subject to the constraints of instructors' schedules. Linear programming and Python are both skills that I have learned—and am continuing to learn—since graduate school (as are many of the data mining techniques that I use in some of my other research). One of the best things about earning a PhD in math or a related field is that it hones your ability to teach yourself, so you can truly be a lifelong learner.

Index

A. Brisbin et al. (eds.), *Research Connections*, Association for Women in Mathematics Series 36, https://doi.org/10.1007/978-3-031-81217-0

MIX
Papier aus verantwortungsvollen Quellen
Paper from responsible sources
FSC® C105338

If you have any concerns about our products,
you can contact us on
ProductSafety@springernature.com

In case Publisher is established outside the EU,
the EU authorized representative is:
Springer Nature Customer Service Center GmbH
Europaplatz 3, 69115 Heidelberg, Germany

Printed by Libri Plureos GmbH
in Hamburg, Germany